ENCYCLOPAEDIA OF STEM CELLS

Vol. VI

STEM CELL CULTURE

By

Dr. Amita Sarkar

Dept. of Zoology
Agra College
Agra (U.P.)
(India)

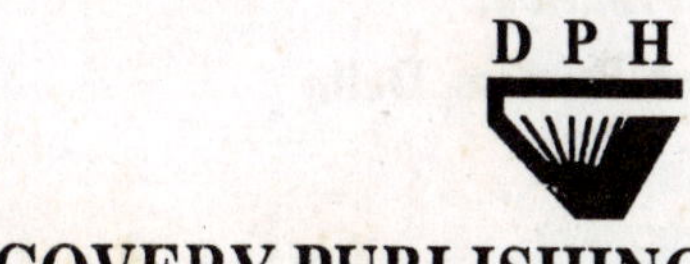

DISCOVERY PUBLISHING HOUSE PVT. LTD.
NEW DELHI-110 002

First Published-2008

ISBN 978-81-8356-358-1 (Set)

Published by:

DISCOVERY PUBLISHING HOUSE PVT. LTD.
4831/24, Ansari Road, Prahlad Street,
Darya Ganj, New Delhi-110002 (India)
Phone: 23279245 • Fax: 91-11-23253475
E-mail: dphbooks@rediffmail.com
dphtemp@indiatimes.com
Website: www.discoverypublishinghouse.com

Printed at:

Sachin Printers, Delhi

Preface

The present title *Encyclopaedia of Stem Cells* is the amazing advancement of biotechnology. It provides the various fundamental aspects of stem cell technologies to be understood adequately. It has been compiled for graduate and undergraduate students, research scholars, teachers, practising biochemical engineers, biotechnologists, applied and industrial microbiologists, cell biologists and scientists involved in bioprocessing research and development. The basic concepts have been clearly explained and their functions are adequately highlighted. The presentation of the text is simple and systematic. The selection of chapters and topics is according to the specified syllabi of several Indian Universities and are so structured as to enable the student to move easily from the fundamental to the complex. It is our earnest hope that this title will be of great value to all our students.

To make the work more comprehensive and informative, the author has consulted many authoritative books, research journals, abstracts, monographs etc. He is grateful to all those great scholars whose work are cited or substantially reproduced.

There can be no claim to originality except in the manner of treatment and much of the information has been obtained from the books and scientific journals available in the different libraries.

The author expresses his thanks to his friends and colleagues whose continue inspirations have initiated him to bring out this book.

The author expresses his gratitude to Mr. Wasan and staff of M/s Discovery Publishing Pvt. Ltd. for their whole hearted co-operation in the publication of this book.

In the mean time, the author will remain sincerely responsible for any shortcomings of the book and the grateful to the readers for their suggestions and constructive criticism for the continuous betterment of the book. He takes this opportunity to appeal to the readers to send their suggestions straightway to his publisher.

Author

CONTENTS

and products, Example 2, Estimation of metabolic quotients in flask cultures, Oxygen, Estimation of ATP production, Example 3, Metabolite yield ratios, Example 4, Calculation of apparent yields without n_v, Cell yield on substrate, ATP maintenance energy model, Immobilized-cell reactors, Estimation of cell density using metabolic quotients, Product Formation, Specific product formation rate, Constant q_P in batch culture, Systems with product retention, Product yield on medium components, Immobilized-cell reactors, Concluding Remarks, Modelling, Background for the Modelling of Mammalian Cell Cultures, Specific rate of cell growth:, Specific rate of cell death, Specific rate of nutrient uptake, Specific rate of metabolite and protein production, Rate of glutamine decomposition, Method for Kinetic Model Construction, Experimental investigations, Example, Kinetic analysis, Identify the rate-limiting nutrients, Identify the rate-limiting metabolites, Calculate the total cell production, Calculate the specific growth rate, Calculate the specific death rate, Model development, Select a rate expression for cellular growth, Select a rate expression for cellular death, Select the rate expressions for nutrient uptake, Select the rate expressions for metabolite and antibody production, Parameter identification, Use of the Model for the Evaluation of Mate-limiting Factors, Calculate the growth rate reduction factors by nutrients and metabolites, Calculate the factors increasing the death rate by nutrients and metabolites, Compare the kinetic effects of the nutrients and metabolites, Cell Death in Culture Systems (Kinetics of Cell Death), Reagents and solutions, Procedures: Morphological Characterization of Cell Death, Materials and equipment, Procedure: Biochemical Characterization of Cell Death, Isolation of DNA, Supplementary Procedure: Purification of Apoptotic Cells, Detoxification of Cell Cultures, Procedure: Detoxification by Dialysis, Materials and equipment, Preparation of dialysis tubing, Detoxification, Alternative Procedure: Detoxification by Gel Filtration, Materials and equipment, Detoxification, Background information, Time considerations, Oxygenation, Units, Effect of oxygen concentration on cells, Typical values of the specific oxygen consumption rate, Procedure; Measurement of Oxygen Transfer Coefficient and Oxygen Uptake Rate, Supplementary Procedure: Oxygenation Methods, Surface aeration, Sparger aeration, Bubble column

1

INTRODUCTION

The last three decades have witnessed a major development in cell culture technology. The field developed substantially and became an essential part of biotechnology. During its evolution, cell culture technology successfully integrated various disciplines, including cell biology, genetic engineering, protein chemistry, genomics, and chemical engineering. Cell culture technology is now the established method of producing a number of important proteins, especially those that are large, complex, and glycosylated.

Cell culture technology-derived products are currently used as medicines to prevent and treat serious diseases such as cancer, viral infections, heredity deficiencies and a variety of chronic diseases. The products are proven to be safe, effective, and economical. The capacity requirements for these products and the market they generated exceed the initial estimates. Some of the cell culture-derived products have a demand of 500 kg/year and generate $1–2 billion in revenue.

Cell culture technology went through a significant evolution from its origin to its commercialization. Cell culture techniques were originated as study tools to investigate cell and tissue behavior and function in vitro. Utilization of cell cultures for therapeutic purposes started with the use of cells for vaccine production. Cultured cells were successfully used as hosts to grow viruses, opening the field of large-scale vaccine manufacturing.

The next big step for cell culture technology was the acceptance of continuous cell lines by regulatory agencies. Continuous cell lines can grow indefinitely, have less stringent growth requirements, and, most importantly, they can be cultured in suspension. Elimination of solid substrate in suspension culture allows scaling up by volume and allows the cells to grow in bioreactors using well-established methods similar to those used for microbial systems.

Genetic engineering and the use of recombinant DNA technology made it possible to produce a vast number of products in cell culture. A number of vectors are now used in a variety of cell lines to produce native and modified human proteins. Engineering the machinery of the cells also allowed the modification of protein products for stability, efficacy, and biological activity. A parallel development in cell culture technology resulted in tissue engineering, where cells are produced as products for tissue replacement and gene therapy.

Advancement and development of cell culture technology required an interdisciplinary approach. As a result of close collaboration and tight integration between cell biology and biochemical engineering, cell lines with excellent productivities can now be cultivated in large-scale bioreactors and at very high cell densities. Issues related to shear sensitivity, aeration, and mixing in bioreactors are largely resolved. The processes can be scaled up successfully and cells can be cultivated consistently in 20,000 L bioreactors. Advances in medium development and cell retention technologies resulted in very high cell densities in the bioreactors. In addition to enhancing productivity in the bioreactors, the cost of the cell culture process was diminished significantly by the elimination of serum and other high-cost proteins from the medium. Now, chemically defined medium is a reality. This progress contributes significantly to the safety and reliability of raw materials in the production of biologicals. Finally, the technology now utilizes efficient separation, purification, and virus inactivation methods that result in highly purified, efficacious, and safe products.

This chapter overviews the development and evolution of cell culture technology from a historical perspective, and reviews the state of the art of cell culture technology. There are several products from cell culture technology approved for treatment of diseases; a brief description of these products is also presented. Finally, a look into the future of the biotechnology field and an assessment of current products in development are provided.

A Brief History of Cell Culture Technology

Methods for growth and maintenance of primary cells in vitro opened the doors for cell culture technology. Early studies with embryos and fibroblasts provided tools for studying cell behavior and function in vitro. These studies also provided a tool for developing cell culture media. The cells used in those days had a limited life span unless they were transformed or were originated from tumors. The production of biologicals from cell culture evolved gradually over the last three decades.

Early Days of Cell Culture Technology—First Products

Utilization of cultured cells for the production of viral vaccines was the first application of cell culture technology. Viruses for vaccine production need living cells to propagate. Embryonic chicken in eggs traditionally was used for vaccine production. Due to increased demand, alternative methods were sought and cell culture technology was the answer. The production of polio vaccine using cells grown in culture started in 1954. The cells were primary monkey kidney cells grown on surfaces (attachment dependent cells). A similar development took place for the vaccine against foot-and-mouth disease for veterinary use. Large-scale vaccine production for FMD virus was a major activity 36 years ago. Increased demand and process economics required more effective and scalable processes. Baby hamster kidney (BHK) cells were adopted for this process and FMD vaccine could then be produced at a 5000 L scale in suspension bioreactors. The medium used for this production was Eagle's medium with 5% adult bovine serum. Virus production continued to utilize cell culture technology, and vaccines such as measles, rabies, mumps, rubella, and varicella are produced in large quantities. In addition to these human vaccines several veterinary vaccines are manufactured by cell culture.

Production of biologicals for medical applications also underwent a significant evolution. Interferon (IFN) alpha was the first cell culture derived product used as a drug. IFNs were traditionally made from human white blood cells from blood donors. The production of IFN was very limited, as it was dependent on donor blood cells and was present at very low concentrations. Work at Wellcome Research Laboratories was instrumental in this area. Initially, researchers at Wellcome wanted to use BHK cells as they could be expanded in suspension and these cells have already been approved for vaccine production. However, IFN-alpha was to be used as a medicine and issues were raised for the BHK cell line. BHK cells contain C-type virus particles and they have oncogenic potential. Nawalwa lymphoblastoid cells, on the other hand, had a better biosafety profile and could be cultivated as easily as BHK cells. The production of IFN moved cell culture technology to the 10,000 L scale and the process is being used successfully for IFN production.

Progress in Cell Line Acceptability

As mentioned before, cells used for early vaccine production were primary cells such as monkey kidney cells and chicken embryo. These cells are attachment-dependent cells requiring a surface on which to grow. In addition the primary cells have stringent growth requirements (growth factors, serum) and their expansion requires a lot of manipulation (trypsinization). These requirements inhibit the use of attachment-dependent cells for large-scale production. The shift from attachment cells to suspension cells was a major advancement for cell culture technology. The use of BHK cells for veterinary vaccines came from the need to meet the demand for FMD vaccine. BHK cells are continuous cell lines and can grow in suspension. Although the BHK cells were accepted for veterinary vaccines, these cells could not be approved for human vaccines. BHK cells were considered unsafe because they contain tumorigenic agents and viruslike particles. As a result, regulatory agencies (WHO, FDA) evaluated other options and, in the end, they approved the use of human diploid cells such as WI-38 and MRC-5. These cells are "normal" cells, virus-free, and contain no carcinogenic substrate. Although the biosafety profile of these cells was acceptable, these cells were still a challenge in large-scale production. They have a limited life span and are attachments-dependent cells.

The concerns of regulatory agencies about the use of transformed cells or cells derived from cancer tissue were addressed by tight monitoring and control of cell lines. In the 1980s, agencies accepted Namalwa cells for the production of IFNs based on characterization data of the master cell bank. These cells did not have any contaminants and the product did not have detectable infective agents. Even though these cells had Epstein-Barr (EB) virus, it was not a serious drawback. The product from these cells could be purified in a process that effectively removes the majority of cell-derived proteins and DNA.

Continuous cell lines such as BHK, Chinese hamster ovary (CHO), myeloma cells (SP2/0, NS0), and human embryonic kidney (HEK) cells were accepted gradually for use in cell culture technology. These cells can be grown in suspension and are easy to scale up. Since the cells are continuous cell lines, their growth requirements are manageable. In fact, these cells can grow in chemically defined media without serum or protein supplements. In addition, these cells can be modified genetically to produce recombinant proteins. In 1987, tissue plasminogen activator (tPA) was approved from recombinant CHO cells. In the last decade, tissue engineering was used to generate tissues for replacement therapy and the cells themselves were approved as products for tissue replacements. The latest development in tissue engineering is the use of embryonic stem cells. These cells can differentiate to different tissues that can be implanted or used as tissue replacement.

Progress in Large-Scale Cell Culture Technology

While small-scale cell culture provided sufficient products for investigational and diagnostic purposes, the demand for vaccines and therapeutic products required large-scale production. Some of the cell culture technology-based products such as monoclonal antibodies need to be manufactured at 1000 kg/year. This can only be done in large bioreactors (up to 20,000 L) by efficient processes that produce several kilograms a day.

Early cell lines used for vaccines were attachment-dependent cells. These cells required medium supplemented with serum and surfaces on which to grow. The expansion and propagation of cells required trypsinization. Scale-up of these systems was achieved by adding more surfaces into the bioreactor system. Roller bottles, T-flasks, and disk propagators are still used to scale up these cultures. A modernized version of these systems is the cell cube, which bears a similarity to a multichamber T-flask. Compared to suspension systems, attachment-dependent cell systems are more difficult to scale up because the surface-to-volume ratio gets smaller at larger scales.

The development of microcarrier cultures was a major break through for attachment-dependent cell systems. The microcarriers can be suspended in stirred tanks similar to those used for microbial systems. Modifications to the impellers and aeration systems ensured the success of the bioreactor process. The cells, when inoculated into microcarrier-containing bioreactors, attach and grow on micro-spheres. The cells can be fed by fresh medium and the product can be removed from the bioreactor continuously. Microcarriers are easier to scale up; one can simply put more microcarriers in larger bioreactors. These systems are now used for vaccine production as well as for protein production at scales of 10,000 L or higher. An improvement in the microcarrier culture was the use of macroporous carriers in the bioreactors. These macroporous carriers allow cells to grow inside as well as outside, thus achieving higher densities by the extra room provided.

The acceptance of continuous cell lines made it possible to grow the cells in suspension in large scale bioreactors. Continuous cell lines are mostly suspension cells that can be agitated and aerated in bioreactors. However, culturing cells in bioreactors was not straightforward. The absence of cell membranes and the stringent growth requirement of cells required both the cultivation procedures and bioreactor operating conditions to be customized. Optimal growth conditions in the bioreactor were achieved by implementation of the results from intensive studies on shear, agitation, aeration, and medium optimization. Today, cells can be cultivated at large-scale bioreactors (up to 20,000 L), at high densities (up to 50 million cells/mL), for many months of operation.

High-Efficiency Cell Culture Processes

The efficiency of cell culture processes is measured by the productivity of the cultures, i.e., the amount made per volume per day. This quantity, the volumetric productivity, is the product of the cell density achieved and the productivity of each cell (specific productivity). Volumetric productivity determines the amount made in a certain production period that can be measured as accumulated product concentration in the bioreactor.

Novel transfection, amplification, and selection methods resulted in high-producing cell clones with high specific productivities. Specific production rate is a measure of how much product is secreted from the cell per unit of time. For monoclonal antibodies, the specific production rate is commonly expressed in picograms per cell per day. For commercial production of antibodies, a minimal specific productivity of 10–20 pg/cell/day is expected. Advances in cell line development and optimization of culture conditions can result in specific productivities as high as 40–100 pg/cell/day.

In addition to specific productivity, the volumetric productivity is directly proportional to the cell density in the bioreactor. Cell density attainable in the bioreactor depends on several factors such as cell growth and death rates, composition of the medium, and aeration capacity. In early large-scale bioreactor development studies, the aeration was a major limitation for achieving high cell density. Over the years effective aeration strategies were developed and most of the issues with sparging have been resolved. An optimized design of the sparger (gas flow rate, diameter, etc.) and some media additives (F68, antifoam) can effectively minimize the impact of sparging. Today, the cell density in the bioreactor is mostly dictated by the media composition and media exchange rate.

Mammalian cells require a highly complicated medium and a delicate formulation is required to maximize cell growth. If the aeration is not limiting, achievable cell density in the bioreactor is controlled by: (a) the nutrient levels in the medium, and (b) the medium exchange rate. Nutrient levels in the medium can be increased by extensive medium optimization. Media development and optimization is an art and a science. Many variables are involved in cell metabolism and the components of the medium interact with each other. Early media formulations provided cell densities of 1–2 million cells/mL. Today, a cell density of 10 million cells/mL is achievable in a batch culture. This is an order of magnitude improvement, accomplished in the last two decades. In perfusion cultures, the cell density can be even higher due to high medium exchange rates resulting in cell densities of 20–50 million cells/mL.

Media Development

Process economics and implementation of cost-effective processes benefited not only by increasing the productivity, but also by reducing the manufacturing cost. Medium development contributed greatly to the process economics by allowing cells to grow to high densities and by decreasing the cost of the medium components.

In the beginning, cell culture medium contained animal products such as serum, albumin, and growth factors. For attachment-dependent cells, other components were also added to stimulate attachment and to form extracellular matrices. These components added to the cost of production and complicated the material sourcing and material release processes. Over the years, serum and other animal products were eliminated from the medium and most continuous cell lines can now be grown in serum- and protein-free media formulations. The use of simple media made it possible to produce kilogram amounts of protein per year economically from volumes close to 100,000 L.

Serum-free media development was followed by the complete elimination of animal proteins from the medium. Initially, albumin and other proteins have been used in serum-free media to substitute for some of the serum components. However, concerns about viruses and animal disease such as bovine spongiform encephalopa-thy (BSE) required the development of animal product-free (APF) media formulations. Early

APF media relied heavily on the use of plant hydrolysates such as soy peptone and yeast extracts for optimal results. Thanks to extensive media development efforts, these components could also be eliminated from the medium with minimal loss in productivity. These efforts resulted in chemically defined media (CDM) formulations. Chemically defined medium is now utilized in many biotechnology processes as the medium of choice because every component could be traced and consistently manufactured.

Products from Cell Culture Technology

Viral Vaccines

Vaccination for the prevention of infectious diseases has been an effective strategy for many years. Polio vaccine was the first cell culture technology-based vaccine and was produced in cultured monkey kidney cells. The cell-based vaccine technology evolved in the last four decades; the production of vaccines now utilizes primary cells, human diploid cells, and continuous or even recombi-nant cell lines. Vaccines against hepatitis B, measles and mumps, rubella, rabies, and FMD had been very effective in preventing life-threatening diseases. New vaccines target human immunodeficiency virus (HIV), herpes simplex virus, respiratory syncytial virus, cytomegal virus (CMV), and influenza, and continue to utilize cell culture technology for production. The latest developments in vaccine development include genetically engineered vaccines and DNA vaccines that will open new frontiers in this field. Development of new vaccines for HIV and cancer is very exciting and these efforts should come to successful conclusion in the next decade.

Cytokines (Interferons and Interleukins)

Cytokines are soluble mediators or glycoproteins helping cells communicate and function as part of the immunological, hematological, and neurological systems. Interferons and interleukins are cytokines with enormous therapeutic potential. Alpha interferon (Wellferon) was developed as the first cell culture-derived biological for treatment of cancer. This was achieved in 8000 L bioreactors using Nawalwa cells at the Wellcome facility. It was followed by the production of several other interferons and interleukins. Due to the small size and relatively simple molecular structure of cytokines, production by simpler cell systems (microbial or yeast) seemed to be adequate. However, mammalian cells express more native cytokines, and cell culture technology is the method of choice for alpha, beta, and gamma interferons. In addition, interleukins 2 to 4, 6, 11, and 12 are also produced by cell culture technology.

Hematopoietic Growth Factors

The formation and differentiation of hematopoietic cells to give rise to mature blood cells require a series of growth factors. These growth factors are mainly single-chain polypeptides and can be produced by microbial systems. However, cell culture technology can offer advantages when the molecule is complicated by glycosylation and when the native form of the molecule is required for therapy.

Erythropoietin (EPO) is a hormone that controls the maturation of red blood cells and it is used for clinical applications in anemia. Recombinant EPO was genetically engineered in CHO cells and the product was launched in 1989 under the names of Epogen (Amgen), Procrit (Johnson and Johnson), and Eprex (Johnson and Johnson). EPO was launched in 1990 in Europe and Japan under the names of Epogin and Recormon.

Growth Hormones

Although conventionally expressed in *Escherichia coli*, human growth hormone can be produced efficiently using cell culture technology. Seostim and Saizen are produced by C127 cells and marketed by Serono S.A.

Monoclonal Antibodies

Antibodies were hailed as "magic bullets" for targeting and neutralizing their antigens as therapeutic agents. Initial application of antibodies involved in vivo and in vitro small-scale production for diagnostic kits. It took almost two decades since then for the field to mature and produce antibodies that are safe and effective therapeutics. Today, antibodies constitute more than 25% of total biotechnology production.

The first antibody was produced by murine ascites. OKT3 antibody was approved in 1987 for the treatment of transplant rejection. Since then both the molecular structure of the antibodies and their production methods evolved significantly. The mouse antibodies were replaced first by chimeric, then by humanized, and recently by fully human antibodies. This allowed the gradual elimination of immunigenicity of antibodies. Antibodies are now produced in hundreds of kilograms using stirred tank bioreactors.

Table 1.1 presents the antibodies approved to date. Antibodies are used for HIV, cancer, allergic diseases, arthritis, renal prophylaxis, septic shock, transplantation, asthma, CMV, and anti-idiotype vaccines. In addition to those presented in Table 1.1, there are a large number of antibodies in clinical trials.

Recombinant Thrombolytic Agents

Recombinant tissue plasminogen activator (tPA) was the first product from cell culture. Genentech obtained the approval for tPA production from CHO cells in 1987 (Table 1.2). tPA is produced in a large-scale (>10,000L) cell culture process. Although the current tPA is highly efficacious, there are several attempts to improve the pharmacokinetics of the molecule. In addition, there is a new generation of recombinant thrombolytic agents under development.

Recombinant tPA from Genentech is marketed as Activase and it is used for acute myocardial infarction, acute massive pulmonary embolism, acute myocardial infarction accelerated infusion, and ischemic stroke within 3–5 hr of symptom onset.

TABLE 1.1 APPROVED MONOCLONAL ANTIBODIES

Product	Company	Indication	Approved
ORTHOCLONE OKT 3	Ortho Biotech	Kidney, heart, and liver transplantation	1991
OncoScint CR/OV	CYTOGEN	Diagnosis of colorectal and ovarian cancers	1992
ReoPro	Centocor/Eli Lilly	Transluminal coronary angioplasty	1994
CEA-Scan	Immunomedics	Diagnosis of	1996
ProstaScint	CYTOGEN	Diagnosis of prostate adenocarcinoma	1996
Verluma	DuPont Merck	Diagnosis of small-cell lung cancer	1996
Neumega	Wyeth	Thrombocytopenia	1997
Rituxan	Genentech/IDEC	Non-Hodgkin's lymphoma	1997
Zenapax	Roche/Protein Design	Kidney transplant rejection	1997

TABLE 1.1 (*CONTINUED*)

Product	Company	Indication	Approved
Thmoglobulin	SangStat	Kidney transplant rejection	1998
Herceptin	Genentech	Breast cancer	1998
Remicade	Centocor	Crohn's disease, rheumatoid arthritis	1998
Simulect	Novartis	Renal transplant rejection	1998
Synagis	MedImmune	Respiratory synctial virus	1998
Mylotarg	Wyeth	Acute myeloid leukemia (AML)	2000
Campath	Berlex/ILEX Oncology	B-cell chronic lymphocytic leukemia (B-CLL)	2001
Zevalin	IDEC Pharm	Non-Hodgkin's lymphoma	2002
Xolair	Genentech	Astma	2003
Bexxar	Corixa	Non-Hodgkin's lymphoma	2003
Raptiva	Genentech	Psiorasis	2003
Tysabri	Biogen/DEC/Elan	Multiple sclerosis	2004
Erbutux	Imclone	Colorectal cancer	2004
Humira	Abbott	Rheumatoid arthritis	2004
Avastin	Genentech	Colorectal cancer	2004

Medicines for myocardial infarction now include TNKase2 from Genentech, which was approved in 2000 and is produced from CHO cells.

Recombinant Coagulation Factors

Blood coagulation factors traditionally obtained from plasma fractionation were used for the treatment of genetic diseases such as hemophilia. The issues related to blood safety, HIV, and Jacob-Creutzfeld Disease required a switch to recombinant protein production using cell culture technology.

The treatment of hemophilia benefited greatly with the introduction of recombinant factor VIII in 1992 (Table 1.3). Baxter/Wyeth obtained the license for rFVIII product Recombinate™ for the treatment of hemophilia A. This was followed by Bayer (KoGENate and KoGENate-FS, approved in 1993 and 2000, respectively) and Wyeth (ReFacto,

TABLE 1.2 RECOMBINANT THROMBOLYTIC AGENTS PRODUCED BY CELL CULTURE

Product	Company	Indication	Approved
Activase	Genentech	Myocardial infarction, ischemic stroke	1987
TNKase	Genentech	Myocardial infarction	2000

TABLE 1.3 RECOMBINANT COAGULATION FACTORS APPROVED

Protein	Medicine	Company	Indication	Approved
rFVIII	Recombinate	Baxter/Wyeth	Hemophilia A	1992
rFVIII	KoGENate	Bayer	Hemophilia A	1993
rFIX	BeneFIX	Wyeth	Hemophilia B	1997
rFVII	NovoSeven	Novo Nordisk	Hemophilia A/B	1999
rFVIII	Helixate FS	Aventis	Hemophilia A	2000
rFVIII	KoGENate-FS	Bayer	Hemophilia A	2000
rFVIII	ReFacto	Wyeth	Hemophilia A	2000

approved in 2000). Other companies such as Aventis market rFVIII under the name of Helixate and Helixate FS.

Recombinant factors VII and IX are now also produced by cell culture technology. Wyeth developed rFIX under the name of BeneFIX™ using CHO cells. This product was approved for the treatment of hemophilia B in 1997. Recombinant factor VII NovoSeven is a product from Novo Nordisk for treatment of bleeding episodes in hemophilia A or B patients with inhibitors to factor VIII or factor IX.

Recombinant Enzymes

CHO cells are used in the manufacture of DNAse. This drug is sold under the name of Pulmozyme and used for the treatment of cystic fibrosis. Genentech obtained the license for manufacturing of Pulmozyme in 1993. Genzyme now manufactures another enzyme, glucocerebrosidase, using CHO cells under the name Cerezyme. This product obtained the license in 1994 for the treatment of Gaucher disease and replaced placenta-derived Ceredase.

Cell Therapy

The latest development in cell and tissue engineering resulted in therapies for replacement, repair, or enhancement of damaged tissue.

The first tissue therapy product was for burn patients. Keratinocytes could be grown in vitro into large sheets of skin tissue that could be grafted. Dermagraft-TC has been approved as a temporary wound covering for partial-thickness burns. Dermagraft-TC was first approved for marketing in 1997 as a temporary wound covering for severe burns (Table 1.4). Dermagraft-TC was the first human, fibroblast-derived temporary skin substitute for the treatment of partial-thickness burns that has been approved for marketing by the FDA.

Other tissue therapy products were also developed. Carticel™ is used to repair clinically significant, symptomatic cartilaginous defects

TABLE 1.4 CELL THERAPY PRODUCTS APPROVED

Product	Company	Indication	Approved
Dermagraft-TC	Advanced Tissue Sciences	Skin replacement for burn patients	1997
Carticel	Genzyme Tissue Repair	Cartilaginous defects	1997
Apligraf	Novartis	Venous leg ulcers	1998
DACS SC	Denderon	Chemotherapy	1999

of the femoral condyle (medial, lateral, or trochlear) caused by acute or repetitive trauma. Genzyme Tissue Repair obtained approval for this product in 1997. Another cell/tissue therapy product is Apligraf™ manufactured by Novartis Pharmaceuticals. This product was approved in 1998 for the treatment of venous leg ulcers. Finally, DACS™ SC from Dendreon was approved in 1999 for rescue therapy following high-dose chemotherapy.

Growing cells and tissues for cellular therapies have many potential applications including diabetes (islet cells), Parkinson's disease (fetal dopamine cells), cancer (heamopoitic cells, bone marrow), liver disease (hapatocytes), and cartilage damage (chrondocytes). The latest developments in stem cells will eventually allow growing, grafting, and replacing any kind of tissue of the human body.

Gene Therapy

Gene therapy offers a permanent fix for some of the genetic diseases by introducing the gene that is missing by a delivery system. In addition, gene therapy can be used for the treatment of cancer and viral infections. Injection of a gene and its possible integration requires a carefully designed delivery system that is safe and effective. Gene delivery can be performed in vivo, in situ, or in vitro using several vectors. The production of vectors, mostly engineered viruses, rely on systems similar to vaccine production.

The first genetic therapy targeted ADA-SCID disease in 1990 with some success. Currently, there are several gene therapy protocols executed as clinical trials for cancer, genetic diseases, and AIDS.

Other Products

There are several other products from cell culture technology that must be mentioned here. Recombinant activated protein C is a major product for the treatment of severe sepsis. HEK 293 cells are used for the production of activated protein C.

Follicle stimulating hormone is used for the treatment of infertility. Follistim (Organon) and Gonal (Serano) obtained approval in 1997 and 1998, respectively, for stimulation of ovulation during assisted reproduction. CHO cells are used for the production of both products.

Soluble TNF receptor, Embrel, is used as a fusion protein TNFR (conjugated to antibody-Fc) for the treatment of moderate to severe active rheumatoid arthritis. Following initial approval in 1998, the product was later approved for juvenile rheumatoid arthritis (1999), for disease modification of active rheumatoid arthritis, and for psoriatic arthritis (January 2002). Embrel is produced from CHO cells.

Future Prospects

The future of biotechnology-derived medicines is bright and full of promise. An industrial survey of biotechnology companies revealed approximately 371 medicines under development. These biopharmaceutical therapeutics target about 425 diseases, as presented in Fig. 1.1 Most of the therapeutics are for cancer, with a total of 178 cancer drugs in development by pharmaceutical companies and the National Cancer Institute. There are 47 drugs under development for infectious diseases, 26 for autoimmune diseases, and 21 for AIDS/HIV conditions. These drugs are at different stages of clinical trials and it is likely that some of them will get approved by the FDA. These drugs will be added to the 95 already approved biotechnology-derived products in the growing assenal against diseases.

The drugs under development are presented in Fig. 1.2, categorized by product type. The majority of the drugs are vaccines (a total of 98) and monoclonal antibodies (a total of 75). There will be more medicines using recombinant human proteins, interferons and interleukins, growth factors, and recombinant soluble receptors. New categories of products such as angiogenesis inhibitors and immune-based therapy are on the horizon. In addition, there are 16 gene therapy protocols under development. Tissue engineering and cell therapy account for 13 trials.

The human genome project and sequencing the human gene will open new frontiers for biotechnology. Some researchers believe the human genome will result in instructions for recreating as many as 300,000 different proteins. This will help researchers increase the number of disease targets from 500 to 10,000. It is the hope of pharmaceutical companies that increased understanding of the human genome will cause a quantum leap in drug development, increasing not only the number of drug candidates but also the effectiveness of the drug, and the success rate in clinical trials. New developments in

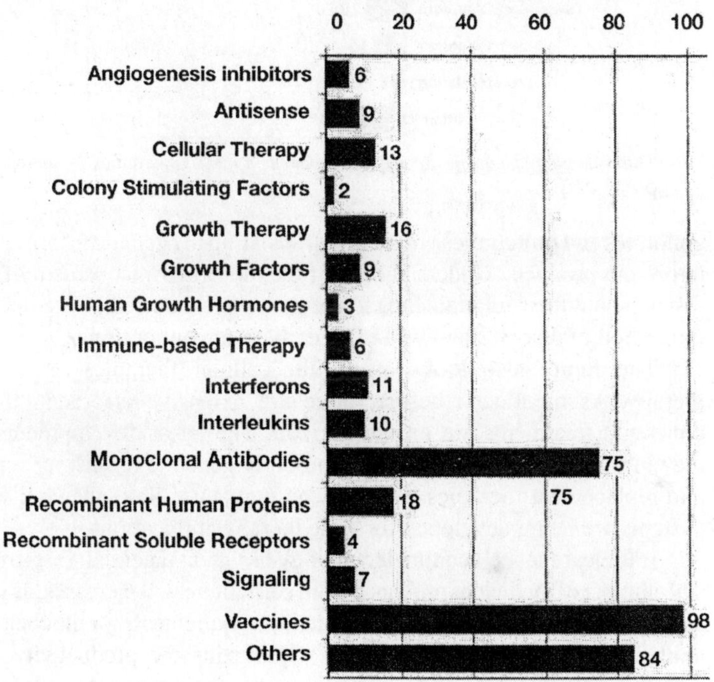

Fig. 1.1 Biotechnology medicines under development in the United States by product category.

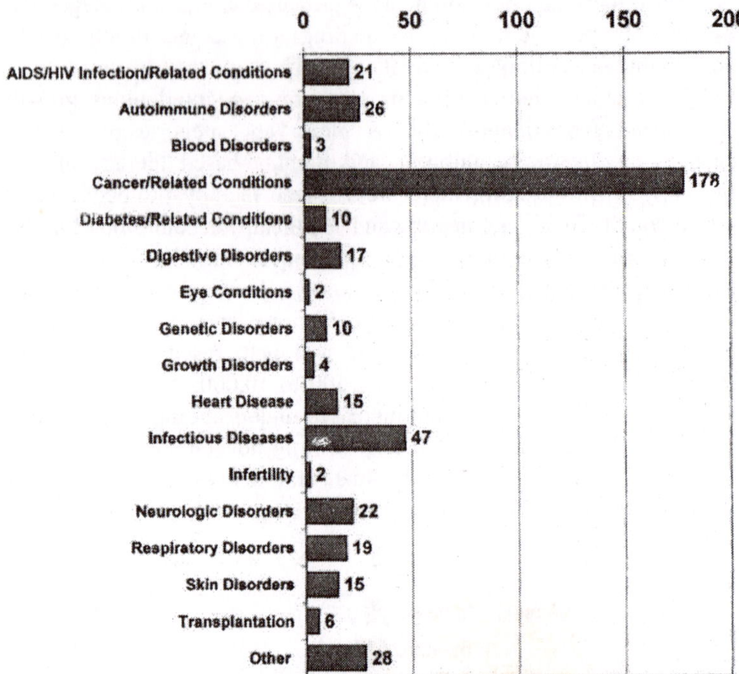

Fig. 2 Biotechnology medicines under development in the United States by therapeutic category.

genomics and proteomics will help in translating the genetic information to protein products. Understanding of the molecular mechanism of the diseases and their relationships to the genes will be used to predict the likelihood of diseases and will offer early treatment options.

The future also looks bright for cellular therapies and gene therapy. As mentioned before, there are extensive efforts in these areas and treatments are on the horizon. The latest developments in the embryonic stem cell area could possibly generate tissues for repair and replacement therapies in the upcoming years. New gene delivery systems are being developed for gene therapy applications.

It is clear that cell culture technology has great potential for growth. The number of medicines produced from cell culture will increases, as well as the amounts to be produced. The potential requirements for monoclonal antibodies particularly will be huge. Optimizing the productivity and minimizing the cost of goods for these products will make it possible for the medicines to penetrate a larger market. The promise of biotechnology is fulfilled and; cell culture technology is here to stay.

2

BIO INSTRUMENTATION

Cultivation of animal cells in an environment optimal for manufacture of desired products requires monitoring and control of a substantial number of physical and chemical parameters. Physical parameters include temperature, fluid flow, and agitation rates. Chemical parameters include the dissolved oxygen (DO) concentration, pH, cell density, and the concentrations of nutrients, cofactors, growth factors, desired products, and wastes. Alterations in the levels of these factors can impact the productivity of a cultivation. Neglecting any of these can potentially introduce substantial variations in the quantity or quality of the desired product. Development of instrumentation to monitor these parameters and to provide aggressive process control schemes has been the subject of substantial research in recent years and now forms a core competency of cell culture technology. This chapter provides an overview of instrumentation applied to monitor cell cultures and describes the application of instrumentation to process control.

Tremendous progress has recently been made in characterizing the influence of the cellular environment on physiology, particularly in identifying and quantifying factors that adversely affect cell growth, viability, and productivity. Achieving high cell concentrations and high viability continue to be common themes in bioprocess engineering research. To reach these goals requires analytical process monitoring schemes and suitable control methodologies to best utilize this information. Use of online measurements to monitor bioprocesses for identifying the physiological state of cells and controlling environmental conditions to maintain the cells in the optimal physiological state has become substantially more prevalent in the past 10 years. Novel control strategies such as expert systems and artificial neural networks have been introduced and applied in bioprocess monitoring and control. While tremendous progress has been achieved in recent years, substantial challenges remain for the full integration of process monitoring with process control. Specific research areas of critical importance for online monitoring and control include: production of genetically modified cultures, production of monoclonal antibodies (MAbs) by cell lines (where control of key nutrients is essential to maintain cell viability and productivity), and development of control strategies using adaptive control or neural networks. Vaidyanathan and coworkers provide a critical assessment of recent developments in measurement technologies with an evaluation of their potential application in industrial production or process development.

The growth of cell cultures is regulated by a complex interaction between the external physical and chemical conditions of the liquid environment of the bioreactor and the internal biochemical processes of the cells. The former conditions are under the direct control of the bioprocess engineer whereas the latter can be modified through genetic manipulations. The challenge, therefore, is to regulate the environment so that the "ideal" conditions are obtained for cell growth or production. For example, hybridoma cells can be highly sensitive to the concentration of available nutrients including glucose. In a batch culture, either excess glucose or a shortage of glucose can diminish the amount of antibody obtained, as demonstrated in the model predictions of Fig. 2.1. This simple model is based on Michaelis–Menten relations to describe the uptake of metabolites, and Monod kinetics with inhibition to describe cell growth. Hybridomas have a clear optimum in glucose requirements to maximize productivity. Unfortunately, identification of optimal bioprocess parameters can be difficult as optima typically are dependent on the values of many other process parameters.

Most traditional efforts in bioprocess monitoring apply well-established techniques to quantify physical factors such as temperature, pressure, agitation, power input, and flow rates and to quantify chemical factors such as pH or oxygen and carbon dioxide in the exit gas. More recently techniques for determining redox potential, DO, and nutrient concentrations have become more prevalent. The area in greatest need for instrumentation development is for biological factors: cell density, cell viability, nutrient concentrations (glucose and glutamine), waste

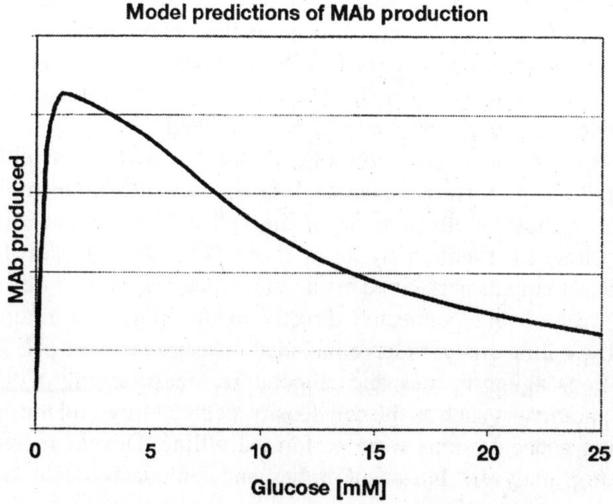

Fig.2.1 Impact of varying glucose concentration on monoclonal antibody (MAb) produced by a batch hybridoma cell culture. Model predictions based on cell growth, death, and metabolism accounting for variations of glucose concentrations. Note that a fairly narrow range of glucose concentrations around 1 mM yield maximal MAb production.

concentrations (lactate and ammonia), and desired product concentrations. Direct measurement schemes are being developed but in general have so far seen only limited adoption partly due to perceived low reliability. Many of these techniques will be described in this chapter.

A complete process control scheme involves: (a) measurement of a response variable, (b) comparison of this variable to set points, and (c) activation of a control scheme. An ideal scheme for monitoring a bioreactor must include an analytical device (a sensor), an appropriate sampling methodology, and a control system to utilize the information gained. The measurements should be rapid and noninvasive; invasive or direct sampling methods run the risk of introducing contamination. A bioreactor with a high degree of monitoring and control systems is presented in Fig. 2.2. Note the interconnectivity of monitoring and process control components.

Monitoring is an important exercise in the operation of bioprocesses and is a subject that has attracted intense research activity in recent years. In order to achieve optimal production or conversion, the factors that influence the performance of a bioprocess should be measured, preferably online, so that the process can be monitored in real-time and appropriate control or remedial action implemented. Such measurement can be achieved either in situ or ex situ, with in situ approaches desirable. In cases where direct analytical information is not available, inferential approaches can be adopted.

Analysis of the cell environment can be performed either "online" or "offline". Here we define "offline" as a measurement that is performed substantially outside of the cell cultivation and often requiring direct intervention by an operator. "Online" analyses then include measurements performed inside a bioreactor (in situ) or outside of the bioreactor, but connected directly to the bioreactor interior. Historically, online analyses have included measurements of pH, DO concentrations, agitation rates, and temperature. Measurements of other chemical parameters such as the cell density, cell viability, and nutrient and product concentrations were performed offline. Development of flow injection analyzers, biosensor probes, and noninvasive techniques

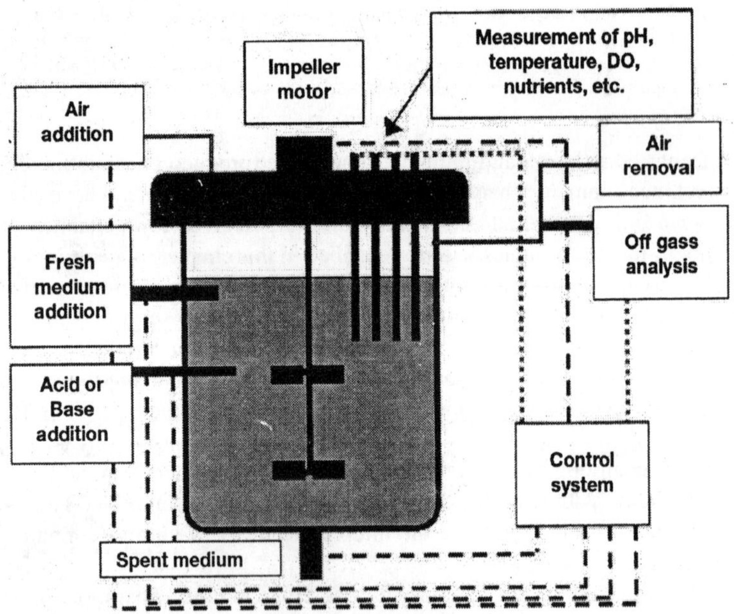

Fig. 2.2 A highly monitored and controlled bioreactor for animal cell culture. Note that all monitoring and process actions follow from decisions made by the control system.

has opened new possibilities for performing more analyses in the preferable online manner.

The function of a sensor is to supply information on the state of the biological process with the ultimate goal of translating this physical or chemical effect into an electrical signal that can be amplified, recorded, and analyzed. For implementation of a control scheme, it is critical that this information reflect the current or future state of the process. Therefore, process sensing must be performed rapidly and preferably with minimal operator intervention.

A bioprocess is significantly different from a standard synthetic chemical reaction in many ways that impact the required instrumentation. For a comparison, Fig. 2.3 presents schematics of a chemical sensor and a biochemical sensor. Biological reactions tend to be stable compared with many chemical processes and most major variations in a biological reaction occur over long time scales (hours to days) in the absence of equipment failures. However, if the cultivated cells have been subjected to an acute stress, they may not be able to recover and the desired product yield may not be obtained.

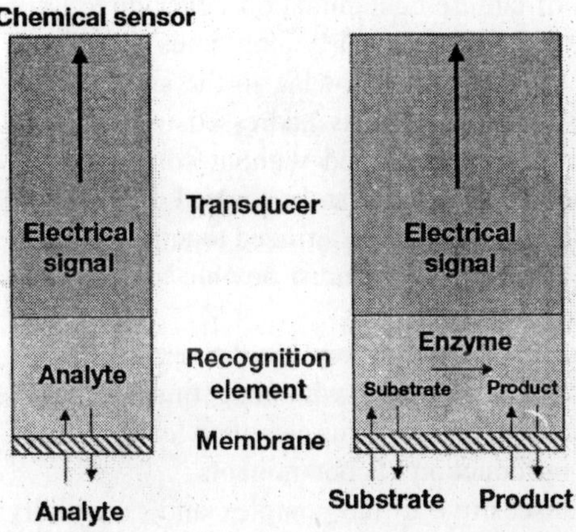

Fig. 2.3 Chemical and biochemical sensors. Both contain recognition elements, transducers, and usually are covered by membranes. Note that the biochemical sensor relies upon a biological reaction (enzymatic or some binding event) to convert the substrate into another component that can more easily be quantified. Chemical sensors do not require such conversions for the analytes to be readily quantified.

Another difference between sensors for a chemical process compared with those for a biological process lies in the complex milieu in which cells grow. The culture media contains proteins, amino acids, sugars, cell debris, and salts. Together these compounds tend to coat or foul surfaces of sensing elements. The methods used to develop sensing technologies for a bioprocess also can depend on whether the cell cultures are anchorage independent (or suspension) or are anchorage dependent, requiring a surface upon which to grow. As anchorage-dependent cells strongly attach to many surfaces, they tend to foul sensing elements much more readily than do anchorage-independent cells. Appropriate, nonsticky, materials that discourage cell attachment must be selected for biosensor components, although some degree of fouling is likely for any surface placed into a bioreactor environment. displays an example of the interconnectivity between cells attached to a surface. Attachment of one cell frequently encourages the attachment of further cells. Due to the biological nature of the contents of cell culture reactors, sensors must meet several strict requirements:

1. The most important factor is that the presence of the sensor must not increase the risk of culture contamination either directly or indirectly. This is critical due to the fairly long times required to establish productive culture systems owing to the slow growth rates of animal cells. This problem is addressed in a number of ways. The sensor can be employed without contacting the interior of the bioreactor; the sensor can be located downstream of the bioreactor; or the sensor can be sterilized together with the bioreactor (in which case the sensor must be able to withstand harsh sterilization conditions).
2. To minimize the need to remove and replace the sensor, it must have a robust and reliable response with long times between failures. Replacement of sensor elements can often lead to loss of process control or to introduction of contaminants.
3. The sensor must be insensitive to the complex and potentially harsh environment of a bioreactor. In many cases, multiple phases are present (liquid, gas, biomass, and suspended bubbles). Additionally, the presence of proteins from serum, cell debris, or desired products must not influence the measurement.
4. The sensor must be highly specific to the desired variable to be quantified. Cell culture media contains many sugars, amino

acids, and small molecules. The presence of structurally similar compounds should not con found measurement.

Sizeable differences exist between the use of monitoring and control methodologies applied in an academic setting (where capital costs dominate decisions) and those in an industrial environment (where time and productivity can be more critical issues). Olsson and coworkers have analyzed the most frequently used bioprocess monitoring components and analytical techniques, with an emphasis on these differences.

MONITORING AND CONTROL OF CELL ENVIRONMENT

Temperature

Temperature is one of the most critical variables to be monitored for the maintenance of healthy cell cultures and so instrumentation and control strategies are well developed and highly accurate. An accuracy of ±0.5°C or less is typically considered adequate for cell culture monitoring, while variations on temperature of 1–1.5° C can in many cases lead to an unacceptable level of production.

The most commonly applied temperature measurement devices for biological processes are resistance temperature devices (RTDs). RTDs have high accuracy and reproducibility with moderately high cost and response time constants of several seconds. RTDs quantify solution temperature based on the changing resistance of a metal conductor with temperature. Platinum is ideal due to its high linearity and broad applicable temperature range and hence is the most commonly applied material.

Thermocouples are also used to quantify temperature in bioprocesses. Thermocouples are lower in cost and more rugged than RTDs; however, temperature measurement is not as accurate or as stable as that obtained from RTDs. For these reasons, thermocouples are usually employed for temperature measurements of processes that are less sensitive to temperature fluctuations or for physically rough environments.

Regardless of the type of temperature measurement employed, this sensing element should be physically separated from the bioreactor interior. For small bioreactors, the thermosensing device is often placed in a deep well that typically runs from the bioreactor head plate downward. This arrangement shields the sensor from corrosion and direct impingement from material flow and allows easy removal and maintenance of the device without the need to shut down the bioreactor. It is critical that both the thermowell and the temperature sensing device be adequately shielded from exterior (i.e., outside of the bioreactor) environmental temperature fluctuations. The thermowell must not be so long that it impacts agitation of the culture medium or be subjected to high shear that may lead to mechanical fatigue.

For use in control schemes, the signal from the sensing device is first amplified, linearized, and then transmitted to a controller, where it is compared to the set-point value. Control of bioreactor temperature is often achieved by regulating the temperature or flow rate of water in an external water jacket or in internal heating or cooling coils. Differences in the temperature of this heat transfer fluid and the culture medium alter the rate of heat transfer.

Simple on-off control usually is adequate for laboratory-scale bioreactors, but better control can be achieved by use of PI (proportional-integral) or PID (proportional-integral-derivative) control schemes. A two-way heating and cooling action using PID control can provide ±0.2°C control on small bioreactors. For larger bioreactors, full PID control using both steam and cooling water are required to provide adequate control. Internal coils including hollow baffles or external heat exchangers may affect heat exchange. Alterations in the agitation rate may also impact solution temperatures.

Flow Measurements

The flow of materials into and out of a bioreactor must be accurately measured for fed batch or continuous flow bioreactors. Numerous types of flow metering devices for sterile or nonsterile solutions are commercially available with the most common being rotameters. These are flow-through devices in which a fluid enters the meter, its force acts upon a float, the float rises within a sight glass with calibrated gradations, and the position of the float can be read by an operator. While rotatmeters are simple to install and to apply, they are not amenable to automation.

Magnetic meters are ideal for monitoring of sterile process fluids, as long as the fluid has some electrical conductivity. The flow of a conducting fluid through a magnetic field generates a voltage that is a linear function of velocity. This voltage is an ideal output for data logging.

Venturi and orifice meters are also commonly used to monitor flow into and out of bioreactors. Unfortunately, they provide low accuracy and are often difficult to clean and sterilize. Therefore, their use has been somewhat limited in bioreactors.

Dissolved Oxygen

Animal cell cultures require oxygen for the production of energy from organic carbon sources. Oxygen can be a limiting species for animal cell cultivations due to its low solubility in water and its high rate of consumption by the cells. The solubility of oxygen in culture media is ~6.6μg/mL at 37°C. Note that this refers to the saturation concentration of oxygen in culture media in contact with a 5% CO_2 95% air gas mixture (standard for cell cultivations). The specific oxygen consumption rate can be cell line dependent. For example, at 37°C, these rates can be 0.31 pmol/cell hr for Chinese hamster ovary (CHO) cells, 0.30 for BHK cells, and 0.22 for hybridoma cells, although specific values will depend on a number of factors including the cell growth rate, carbon source, etc.

Due to variations in cell concentrations and the high consumption rate, the amount of oxygen dissolved in the liquid culture medium is in a state of dynamic equilibrium. At a constant temperature, the amount of DO follows Henry's law:

$$C_L = HC_G$$

where C_L is the concentration of DO in the culture media, C_G is the concentration of oxygen in the vapor phase above the media, and His Henry's law constant. Henry's law constant varies with temperature and composition of the medium.

The most commonly applied means for quantifying DO is the Clark-type electrode, developed in the 1950s by Leland Clark. These consist of an electrode covered by a membrane that is selectively permeable to oxygen. Oxygen diffuses across the membrane and is reduced at a noble metal cathode that is negatively polarized with respect to a reference anode. Polarographic electrodes require that an external voltage be applied for negative polarization and oxygen reduction; galvanic electrodes rely on a voltage generated from the use of electrons from lead, zinc, or cadmium anode.

Dissolved oxygen is most commonly quantified amperometrically, by the reduction of oxygen at the cathode and the formation of silver chloride at the anode. An electrolyte solution connects the anode and cathode, and with a polarizing voltage. Oxygen from a fluid diffuses across a gas permeable membrane and reacts with two water molecules and incorporates four electrons to yield four hydroxyl ions.

A polarographic electrode (Clark-type) usually contains a platinum cathode, a silver/silver chloride anode, and an electrolyte such as potassium chloride. When a voltage of −0.6 to −0.8 V is applied to the anode, the following half-cell reactions result:

$$\text{Cathode: } O_2 + 2H_2O + 4e^- \rightarrow 4OH^-$$

$$\text{Anode: } Ag + KCl \rightarrow AgCl + K^+ + e^-$$

$$\text{Net reaction: } O_2 + 2H_2O + 4Ag + 4KCL \rightarrow 4OH^- + 4AgCl + 4K^+$$

The reduction of oxygen produces a voltage dependent current that is directly proportional to the oxygen activity in solution. The rate of oxygen diffusion through the gas permeable membrane is typically the limiting step.

A galvanic (potentiometric) electrode with a silver cathode and a lead anode has the following half-cell reactions:

$$\text{Cathode: } O_2 + 2H_2O + 4e^- \rightarrow 4OH^-$$

$$\text{Anode: } Pb \rightarrow Pb^{2+} + 2e^-$$

$$\text{Net reaction: } O_2 + 2H_2O + 2Pb \rightarrow 2Pb(OH)_2$$

By carefully selecting the anode material, the cathode will be charged at −0.6 to −0.8 V with respect to the anode. Galvanic electrodes also have a slow step of oxygen diffusion and so yield a linear relation between current output and DO concentration. One difficulty with amperometric DO probes is that they tend to have a high rate of failure.

A spectrophotometric method can be used for determination of DO concentrations in suboxic water (less than 1.0mg/L DO concentration). The method is based on a Rhodazine-DTM colorimetric technique adapted by White and coworkers, which minimizes atmospheric interaction with the water sampled. This technique is sensitive to

0.2µmol/L (0.006 mg/L)–an order of magnitude lower than the amperometric method and much higher than generally required for cell culture work. The technique was originally developed for quantifying oxygen in environmental applications such as ground water, lakes, and reservoirs.

Oxygen concentrations in bioreactor media have also been quantified using fluorescence quenching of suitable dyes. Blue light (450 nm) is produced following the excitation of a ZnS:Ag phosphor from beta particles released by a self-powered radioluminescent (RL) light source (Pm-147), which excites a ruthenium complex immobilized in a membrane. Analytical information is acquired by measuring the magnitude of oxygen-induced fluorescence quenching of the ruthenium complex. This oxygen sensor shows good stability and reversibility, with detection limits of 0.25 Torr in gaseous samples and 0.028 ppm in aqueous samples. This sensor was recently applied to monitor the oxygen content of fibroblast cell cultures sent on the space shuttle mission STS-93.

Gaseous measurement of oxygen can be applied by taking advantage of the paramagnetic nature of oxygen. A dumbbell-shaped, glass rotor is suspended by a torsion spring in a non uniform magnetic field. Oxygen preferentially accumulates in this field and displaces the nonmagnetic rotor. This rotation is opposed by the torsion spring and occurs in proportion to the oxygen tension. The rotation of the rotor can then be read against a calibrated scale. The partial pressure of oxygen in the exit stream is quantified; however, water vapor must be removed from the gas phase in order to minimize sensor drift.

Carbon Dioxide

The concentration of gases in the vapor phase leaving a bioreactor can be indicative of the cellular metabolism and so is a commonly used measure of activity. Carbon dioxide (CO_2) and oxygen are the primary components of interest. At high viable cell concentrations in large-scale mammalian cell culture processes, dissolved CO_2 can accumulate due to low removal rates at reduced surface-to-volume ratios. High concentrations of dissolved CO_2 can inhibit cell growth and metabolism and can impact the glycosylation of some desired protein products. Bachinger and coworkers used a gas sensor array to detect bacterial contamination in CHO cell cultures. Measurement of head space composition provides a means to detect contamination earlier than conventional methods.

CO_2 in the gas phase is commonly quantified using relatively inexpensive infrared (IR) analyzers and are available with many incubators and bioreactors. These sensors are generally stable as long as the ambient temperature is maintained constant and the vapor environment contains a consistent level of moisture. Both of these factors are controlled to maintain cell viability and so CO_2 sensor failure for these reasons is rare.

Early optical schemes to quantify CO_2 were based on a "wet covered" type sensor, in which a pH-sensitive dye in an aqueous buffer is covered by a gas permeable, ion impermeable, membrane. Eventually, measurement approaches evolved to "solid-water droplet" type sensors and then onto "solid" sensors. Mills and Eaton have reviewed the use of optical sensors for measurement of carbon dioxide, particularly through use of luminescence-based sensors.

Luminescence resonance energy transfer has been used as an alternative optical means to quantify CO_2. The basic principle is radiationless energy transfer from a ruthenium complex as luminescent donor to thymol blue (a common pH indicator) as acceptor. These are both embedded in a hydrophobic matrix. In the presence of CO_2 thymol blue is protonated and changes its color from blue to yellow resulting in a decrease in the rate of energy transfer and consequently an increase in decay time. Other dyes can also be used as acceptors and interfaced with a fiber-optic luminescent sensors.

Some novel schemes have been developed to quantify CO_2 using IR absorption. For example, dissolved CO_2 in mammalian cell cultures has been quantified using in situ fiber-optic sensors. This group applied an YSI 8500 sensor that had a response time of 6 min with a sensitivity of 0.5% and high linearity. These measurements were not affected by culture pH or the presence of varying concentrations of cellular metabolites. Chang and coworkers applied a fluorescence lifetime-based sensing film to quantify CO_2 in cell cultures. Uttamlal and Walt developed a fiber-optic CO_2 sensor using a pH-sensitive dye in a bicarbonate buffer for online monitoring of bioreactors. Zhang and coworkers have developed stabilization methods to remove the effect of temperature variations on IR CO_2 measurements. Recent developments in other fiber-optical chemical sensors have been review by Wolfbeis.

Gas chromatography (GC) can also be applied to monitor exhaust gases including CO_2 and others volatile components. Ethanol, acetaldehyde, and car-boxylic acids can be quantified and used to evaluate the cellular metabolic rate. A limitation of applying GC for bioreactor monitoring is that the method requires intermittent injections, often spaced 15min apart, to monitor transients. Most metabolic changes occur on a longer time scale, but the process is not readily amenable to some control schemes.

Mass spectrometry (MS) can be employed to quantify several components simultaneously with a rapid response time (less than a minute) and high sensitivity (detecting concentrations less than 10 μM). MS is gaining in acceptance for process monitoring, although cost of the instrumentation can still be substantial compared with other monitoring schemes. MS instruments are often multiplexed to monitor multiple bioreactors at a time.

pH

Next to temperature, pH is considered the most critical parameter to monitor in a biological process based on the degree of cellular damage that may ensue upon loss of control. Variations of pH of 0.1 units away from optimal can have a substantial impact on cell growth and productivity for certain cell types. While most cell culture media provide substantial buffering of pH, cellular metabolism invariably decreases pH due to the secretion of lactic acid and CO_2 as metabolic byproducts. Hydrogen ions can have a substantial impact on cell metabolism and growth by altering rates of substrate uptake and product release. Typical animal cell culture media begins at a pH of 7.4, but can quickly drop to below 7 within hours, depending on cellular productivity and culture buffering capacity. Difficulties in controlling pH can be related to the logarithmic relationship between pH and the hydrogen ion concentration.

pH-measuring devices are based on the formation of a galvanic cell with two metal conductors connected by an electrolyte. The metal conductors each form a half-cell, one that serves as the sensing element, while the other serves as the reference. A typical glass pH electrode includes a thin-walled glass membrane of 0.2–0.5 mm (which makes the electrode very delicate) enclosing an internal standard solution, usually composed of aqueous HCl. Into this solution is placed a rod of silver covered with silver chloride which acts as an internal reference. The key to the selectivity of the electrode is the glass membrane. The glass chosen consists of chemically bonded Na_2O and SiO_2 and is low in Al_2O_3 and B_2O_3. The surface layers of the glass consist of fixed silicate groups associated with sodium ions. When this electrode is first placed in water, the sodium ions exchange with the sol-vated protons in the water thus forming a hydrated gel layer 50–5000 A thick. This layer exists on both sides of the membrane and is essential for the operation of the electrode.

The use of the pH-responsive glass is restricted to the bulb-shaped element at the end of the electrode, thus limiting the response to only a small sample area. Fig. 2.4 presents a schematic of the surface of a glass pH probe placed in solutions of varying acidity. A constant potential is maintained at the inner surface of the glass membrane by filling the bulb with a buffered solution of stable and accurate hydrogen ion activity. An $Ag/AgCl_2$ electrode is generally used as the electrical lead out from this system. As the pH of the sample varies, this causes a change in potential on the outer surface of the membrane. To measure this a reference electrode is necessary to complete the circuit. The combined electrode is constructed as an integral part of the electrode

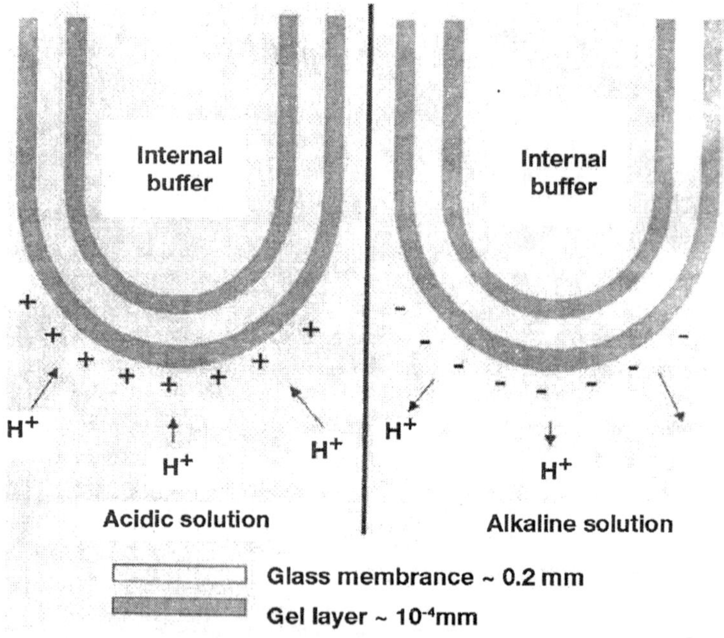

Fig. 2.4 Schematic representation of the tip of a glass pH electrode in two solutions. The external gel layer carries either a positive or negative charge based on the acidity of the surrounding solution.

assembly and consists of an Ag/$AgCl_2$ electrode in KCl electrolyte saturated with $AgCl_2$.

When the electrode is placed in a solution of unknown pH, the activity of the H+ ions in the test solution is likely different than that of the H+ ions in the hydrated layer which sets up a potential difference between the solution and the surface of the membrane. The magnitude of this boundary potential will be determined by the difference in activities. A similar boundary potential will exist at the inner glass wall. The potential can be related to the activity by the Nernst equation:

$$E(\text{cell}) = E^{*} + 0.0591 \log a\,(H^{+})$$

where a (H^+) is the activity of H^+ and E^* includes the standard electrode potential of the glass electrode, the potential of the reference electrode, and liquid junction potentials. pH is then defined as:

$$\text{pH} = -\log a\,(H^{+})$$

Note that the degree of hydration of the outer surface will also change if the electrode is allowed to dry out. It is important that the electrode be stored either in an electrolyte solution (e.g., 4 M KCl) or wrapped with a damp cloth.

The glass electrode generally exhibits a Nernstian response over most of its working range; however, at extremes of pH, the behavior becomes non-Nernstian. Alkaline errors are due to the response of the glass membrane to other ions and so lead to pH measurements that are lower than their true value. These can occur at pHs around 9–10. Acid errors occur at the opposite end of the pH scale and are due to the impact of variations in the activity of water in solution. The net effect is for the sensor to report an acid pH that is lower than the true value.

An alternative pH sensing approach employs a diaphragm that links an electrolyte solution and the sample to be measured. However, sulfur-containing compounds such as the amino acid cysteine can cause these pH probes to fail due to the formation of silver sulfide deposits on the diaphragm surface.

Several important points in the application of pH sensors emerge from the Nernst equation:

1. The millivolt output is positive below a pH of 7 and negative above a pH of 7.

2. The temperature effect at a pH of 7 is 0; however, at pHs away from this point, the effect increases linearly with pH.

Ideally, pH calibration should involve measurement at three points so as to best correct for differences between the Nernst equation and experimental conditions.

Once a probe is calibrated, there are a number of problems that can lead to erroneous or unreliable readings. These include slow response, drift, and reduced sensitivity. Inside of a bioreactor these problems are often the result of:

1. The electrode surface has been coated by proteinatious material from serum, cell debris, or cellular products. Often this will result in a slow mea surement response with pH readings higher or lower than anticipated.
2. The glass surface has been etched or dried. Overly high pH readings and difficulty in calibrating over an extended range may result.
3. Probe terminals are shorted. The probe may report a constant pH of 7 regardless of the true value.

Control of pH is achieved through pulses of acid or base (usually only base is required). A proportional controller can be applied to respond to pHs lower than a set point with addition of a pulse of base (often KOH, NaOH, $NaHCO_2$, or Na_2CO_3). The frequency of measurement and pulse addition and duration of pulse addition can be modified to effect varying levels of control. For pilot plant bioreac-tors, control of pH at ±0.03 pH units is obtainable using pumps operating on a time-proportional basis. Most media applied for animal cell cultivation have substantial buffering capacity around 7.4 providing some degree of safety around the pH desired by cultures.

pH has also been quantified by optical means. For example, the diffuse reflectance of a conductive polymer such as PoAnis/TSA and cellulose acetate fixed on the end of a fiber-optic probe which allows measurements in the pH range from 4.9 to 10.5 with a precision of ±0.01 pH and a response time of 5min for a full pH variation from acid to base. Unfortunately, the sensor response can be a function of the ionic strength of the solution and on the supporting electrolyte used to adjust the ionic strength. Lin reviewed developments of optical and fiber-optic pH sensors in the 1990s including methods for immobilization of pH indicators. Optical and fiber-optic pH sensors are reviewed, including pH sensors based on conductive polymers, imaging fibers, microparticles and nanospheres, as well as micrometer and submicrometer fiber-optic pH sensors, distributed fiber-optic pH sensors, pH sensors for high acidity and alkalinity, pH sensors with broad dynamic range and linear response, and CO_2 and NH_3 sensors based on pH indicators.

While major efforts in cell culture systems are focused on increasing the viable cell density, such increases in cell numbers will also result in an increased oxygen demand and will require better mixing approaches. Additionally, the accompanying increase in CO_2 production and accumulation and the resulting reduction in pH are also important implications for process engineering. Such pH reduction is typically controlled by the addition of sodium carbonate, but with current bioreactor operating schemes, pH excursions away from pH set points can be found in bioreactor regions in close proximity to the location of base addition due to poor local liquid mixing. Enhanced mixing schemes are required to homogenize the bioreactor media without damaging cell function.

Metabolites and Products

Animal cell culture technology has advanced due to increased understanding of cellular metabolism including the effects of product inhibition. Glucose and glutamine are the primary sources of energy and building blocks for animal cells; ammonia and lactate are the primary metabolic byproducts that may influence cellular productivity. As little as 4mM ammonia can reduce the cell growth rate of hybridoma cells by 50% and accelerate metabolic rates in an undesired fashion. Cultivation of hybridoma cells under tightly controlled glucose and glutamine concentrations can increase the cellular production of MAbs. Nutrients and products have been quantified through a variety of methods including electrochemiluminescence, optical sensing, enzymatic conversions, chromatography, and nephelometry.

While monitoring provides important information on the bioprocess, oftentimes, predictive information is also required. This can be achieved through use of biological models. Unstructured models are empirically based on experimental observations. Structured models are theoretically based on knowledge of the individual cellular processes. Miller and coworkers developed an unstructured model for hybridoma growth, death, and MAb production based on measurements of the viable cell density and dilution rate. Dalili and coworkers developed a model to characterize the influence of glutamine on hybridoma growth and MAb production. In agreement with experimental results, Dalili's model predicted that the maximum viable cell concentration increased with glutamine levels in the range of 0.5–2.0 mM. Several groups have developed structured models to characterize the interplay of both nutrients and metabolic byproducts on cell behavior. These models focus on the individual metabolic rates and so provide information which is of most use for development of bioreactor control schemes to optimize MAb production.

Glucose and lactate concentrations have been monitored online using a commercially available analyzer (Model 2700, Yellow Springs instruments, Yellow Springs, OH) during batch and perfusion hybridoma cell cultures. A schematic of this sensing element is presented in Fig. 2.5. Cell-free samples from the reactor were obtained using a 0.45 mm hollow fiber filtering system placed in a circulation loop. A process control strategy was developed to control the concentrations of glucose and lactate in a perfusion reactor by adjusting the feed rate to maintain desired set points. During exponential growth the control algorithm successfully adjusted the perfusion rate while maintaining constant glucose and lactate concentrations. Glucose consumption and lactate accumulation rates were used for the estimation of viable cell density in the reactor.

Recent evidence suggests that there may be a substantial need for monitoring a greater number of components in bioreactors for animal cells than are currently taken into account in such models or are measured online. Simpson and coworkers have demonstrated that depletion of any single amino acid in a hybridoma cell cultivation can lead to apoptotic cell death including both essential and nonessential amino acids. Based on this result, monitoring the glucose and glutamine concentration alone may not suffice for maintaining a healthy culture. Likely all or nearly all of the high-level components in a cultivation must be tracked. Many standard culture media for animal cells contain 17 components present at concentrations of 0.5 mM or greater. Currently, reliable methods are not available to rapidly quantitate the concentrations of all these chemical species in a manner amenable to the implementation of intelligent control schemes. Metabolic engineering of animal cells provide the potential to improve productivity of cell cultures in generating complex proteins with correct post-translational modifications; however, for these approaches to be truly optimized, the

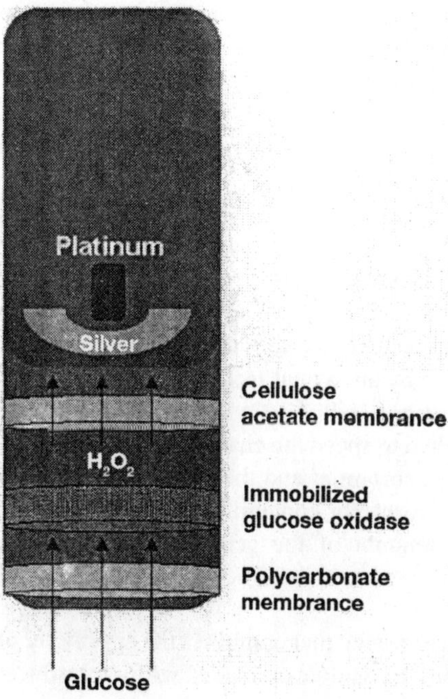

Fig. 2.5 Schematic representation of the sensing mechanism of the YSI glucose analyzer. Glucose diffuses across the polycarbonate membrane, reacts with the immobilized glucose oxi-dase to generate hydrogen peroxide, which diffuses across a cellulose acetate membrane, and is detected by a platinum-silver electrode.

complex requirements for the growth and production must be integrated with an understanding of the cellular environment. Mulchandani and Bassi have recommended that a specific area of importance for the development of online process improvement is the monitoring and control of nutrient concentrations for the maintenance of cell viability and production.

Once a monitoring and/or modeling scheme is developed for an application, the gathered information must be put to use so as to optimize the culture productivity. However, traditional control theory does not take into account the inner metabolic processes that govern the cell behavior such as the interplay between multiple metabolic pathways. Konstantinov presented a methodology for the design of systems capable of performing advanced monitoring and control functions based on the physiological state of a cell population. The physiological state should incorporate the specific rate of metabolism, metabolic ratios for product generation, and the specific growth rate. The goal of the control system then becomes to maintain the physiological state of the cell as close as possible to a predetermined trajectory of maximum efficiency.

Enzymatic Methods

Accurate medium concentration information is required to implement intelligent bioreactor control schemes. For many years, mammalian cell metabolism was determined by removing samples from a bioreactor and analyzed offline by analytical techniques such as high-performance liquid chromatography (HPLC). Online chro-matographic analyses were eventually applied to speed the characterization process, but measurement time can be substantial and the periodic removal of samples creates concerns for bioreactor contamination not to mention the undesired loss of some amount of the generated product. The advent of enzymatic sensors has reduced the time required for analysis of bioreactor media.

A biosensor is an analytical device that combines the specificity of a biological sensing element for the analyte of interest with a transducer to produce a signal proportional to the target analyte concentration. Enzymes have been employed in a wide range of sensing schemes to quantify cellular nutrients, wastes, and products. Most include enzymes from the classes of hydrolases, lyases, and oxidoreductases. Enzymatic biosensor probes provide rapid measurements from within a bioreactor, thus bypassing some of the aforementioned problems. Such enzymatic biosensors can be difficult to sterilize and can be unstable when used over long operation periods because of enzymatic degradation, and, hence, frequent recalibrations are required to ensure accurate measurements. Enzymatic systems also require a separate sensing element for each analyte, complicating measurement of multiple species.

Measurements using enzymatic biosensors are most often performed offline due to the need to isolate the enzyme from the bioreactor environment. Typically in an enzymatic sensing scheme, the analyte to be quantified reacts specifically with an enzyme and generates a product that can be readily detected. In some cases, multiple reaction steps are required. For example, glucose can be quantified by the following reaction scheme:

$$\text{Glucose} + O_2 \rightarrow \text{gluconic acid} + H_2O_2$$

$$\text{Fluorescein} + H_2O_2 \rightarrow \text{fluorescein}^* \text{ (fluorescent)} + 2H_2O$$

The first reaction is catalyzed by glucose oxidase, the second by horseradish peroxidase and is a frequently applied scheme for glucose quantification. The amount of fluorescence detected will correlate to the amount of glucose present, as long as excess oxygen is available. This scheme could also be used to quantify oxygen if sufficient glucose were available. Alternatively, H2O2 could be quantified amperometrically versus an Ag/AgCl reference electrode. Many other similar enzymatic methods have been applied for glucose measurement in biological samples.

Glutamine can be quantified by the following reaction scheme:

$$\text{Glutamine} + O_2 \rightarrow \text{glutamate} + NH_3$$
$$\text{Glutamine} + O_2 \rightarrow \alpha\text{-ketoglutarate} + NH_3 + H_2O_2$$

The first reaction is catalyzed by glutaminase; the second is catalyzed by glutamate oxidase. Such an approach provides multiple options for the ultimate detection of glutamine. H_2O_2 could be measured amperometrically or could be converted by the horseradish peroxidase reaction shown above. The liberated NH_3 could be quantified using an ammonia detection scheme.

When applying enzymes as the biological recognition element in a biosensor, both operational stability and long-term stability must be taken into account. If the enzymes are to be immobilized to a surface (often a requirement for application in which the sensing element is to be separated from the culture medium), the immobilization technique will have a great impact on sensor stability. Chemical methods such as covalent cross-linking to a support material typically yield long stability, but at a reduced enzymatic activity. In some cases, the activity of an immobilized enzyme can be reduced by as much as a factor of 20 compared with a solùble preparation. Physical methods such as entrapment of the enzyme within a membrane or gel material provide greater activity, but at the expense of overall stability of the enzyme and of the sensor as a whole. Culture media proteins, cells, and cell debris can readily foul membranes.

Flow Injection Analysis

Flow injection analysis (FIA) is based on the introduction of a liquid sample into a moving aqueous carrier stream. The injected sample forms a pulse which is transported toward a detector which continuously records the sample absorbance, electrical potential, fluorescence, or other physical parameter that changes as a result of the sample material. Typical output is a broad peak because the sample disperses along the carrier stream. Detection methods include amperometric, potentiometric, fluorescence, chemiluminescence, UV/Vis absorbance, or turbidometric. Advantages of FIA include a reduced risk of contamination, ease of recalibration or replacement of sensing elements, small sample requirements, and short response time. Such methods provide rapid analyses with measurements requiring as little as 20 sec. A disadvantage is that each sensor can monitor only one analyte and that sensitivity and selectivity are often not as high as required for implementation of control strategies. However, multiple sensing elements can be integrated to provide concentration information on a wide range of components.

An FIA system typically includes a selector valve to switch injection flow between a sample to be analyzed and calibration standards; a multichannel pump to deliver sample, reagents, and calibration solutions; an injection valve with a fixed-volume loop (usually 20–50 mL); and a detector. If multiple analytes are to be quantified, the sample flow is often split into different channels, each of which is used to analyze for one component. Isolation of these channels can simplify analysis by providing physical isolation of each sensing array, thus reducing interference or cross-talk between analysis steps. FIA can be difficult to apply in automatic control systems as frequent sampling of culture medium will reduce some of the natural advantages of this technique.

Examples of components in bioreactor media that have been quantified by FIA include ammonia, glucose, glutamine, lactate, and antibodies. Quantification of glucose can proceed through amperometric measurement in which immobilized glucose oxidase produces hydrogen peroxide which can be quantified through anodic oxidation. Quantification of glutamine can be based on measurement of ammonia ions produced in a flow-through enzyme reactor containing immobilized glutaminase enzyme, followed by subsequent downstream potentiometric detection by an ion-selective membrane electrode.

White and coworkers applied three amperometric FIA biosensors based on rhodinized carbon electrodes to quantify glucose, glutamine, and glutamate in parallel in mammalian cell perfusion cultures. The geometry of an FIA system (in which the sensors are not directly incorporated into the bioreactor) permits easy replacement of sensing elements. The inclusion of frequent recalibrations of the sensors maintained accuracy of measurements.

A solid-state, diffuse reflectance-based fiber-optic sensor has been used to quantify ammonia by employing immobilizing chlorophenol red, a weak acid chro-mophoric indicator dye, in a microporous polypropylene membrane. A schematic flow injection scheme is presented. This sensor uses flow-injection to carry a 10-mL aliquot of the sample across the treated membrane. Ammonia in the sample diffuses through the air-filled pores within the membrane structure before reacting with the indicator dye. A reversible acid-base reaction between ammonia and chlorophenol red results in a measurable change in the reflectance at 560 nm. Response characteristics include a peak response within 20 sec, a limit of detection of 0.2 ±0.1 mM ammonia, and a dynamic range of up to 60 mM ammonia.

A flow-injection sensor using an expanded microbed as the enzyme reactor has been developed to quantify glucose and lactate in bioreactors. The expanded bed reactor is capable of handling a mobile phase containing suspended matter like cells and cell debris, thus eliminating the need to pretreat the sample to remove particulate matter. Glucose oxidase and lactate oxidase were immobilized and used to quantify glucose and lactate, respectively, with an assay time of 6min.

The FIA system has also been used as a two-step immunoassay-based determination of human prolactin (hPRL) concentration along with its degree of glycosyla-tion in which the antibodies are immobilized on the surface of a carrier. The results of the two-step FIA method were found to agree with those obtained by the standard methods while requiring minimal analysis time (10min) and elimination of operator intervention.

Schugerl provides an excellent review of recent developments in monitoring product formation processes using FIA, spectroscopic methods, and biosensors. This review also covers advanced control of indirectly evaluated process variables by means of state estimation/observer, with the use of structured and hybrid models, expert systems and pattern recognition for process optimization. Christensen also provides a detailed review on the use of FIA and sequential injection analysis (SIA) for monitoring biological processes.

Spectroscopic Methods

The aforementioned sensing methods are destructive and require that samples be removed from the reactor prior to analysis. The methods also

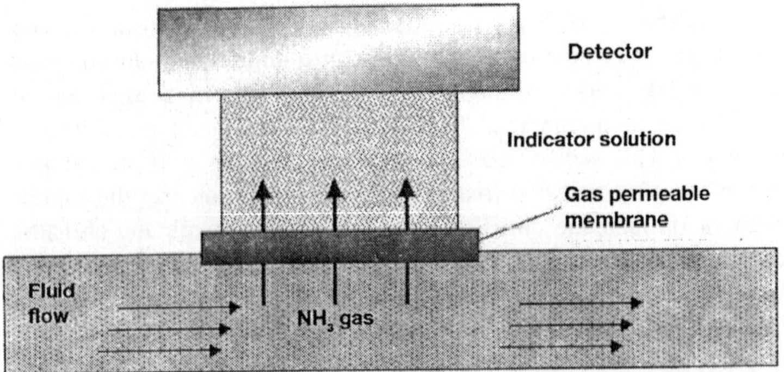

Fig. 2.6 Schematic of an ammonia sensor applied in a flow scheme. NH_3, and not NH_4^+, can traverse the gas-permeable membrane, interact with a suitable indicator solution that may change color, or fluoresce in response to NH_3 that is monitored by the light detector.

typically quantify a small number of analytes, following the axiom of one sensor for one analyte. Efficient bioprocess control requires real-time monitoring of biomass, substrates, intermediates, and nutrients and is performed preferably without destroying or removing bioreactor media. Optical sensors have the advantages that they can be performed quickly, require no sample preparation, and can be noninvasive. Currently, optical sensors are available to quantify DO and dissolved carbon dioxide; however, methods are currently in development to quantify a variety of nutrients and wastes in addition to quantifying biomass.

Spectroscopy using visible, UV, or near-infrared (NIR) light provide attractive alternatives to monitor the concentration of nutrients and wastes in cell culture media. The approach of applying spectroscopy to quantify components in bioreactors has evolved from development of portable glucose sensors designed for use by diabetics. Spectroscopic measurements can be noninvasive, nondestructive, rapid, require no sample preparation, and used to quantify multiple chemical species simultaneously. Spectroscopic measurements are based on introducing a beam of light to a sample, collecting transmitted or reflected light, and correlating the amount of light absorbed to the composition. Any nonsymmetric chemical species may be quantified by identification of its characteristic absorbance features in the NIR. Conversely, symmetric species can readily be quantified using Raman spectroscopy. Many of the critical metabolites, such as glucose, glutamine, ammonia, and lactate have distinct spectral features in the IR, thus suggesting that they may be easily quantified (Fig. 2.7). Fig. 2.8 presents IR spectra of animal cell culture media, demonstrating that in a complex solution, identification of the metabolite features can be difficult.

Each of these measurement types provides specific advantages. FT-NIR provides the greatest light penetration depth and so can be used for thicker samples so that they provide a higher degree of light scattering. FT-mid-IR (MIR) provides information that is more easily discernible as being specific for certain analytes as these wavelengths are closer to the fundamental IR absorptions. FT-Raman is advantageous when the interference due to water is to be minimized, such as when the water content is variable. Sivakesava and coworkers compared these three methods for monitoring a lactic acid fermentation and found that the MIR region was better suited for quantifying glucose, lactic acid, and biomass. Rhiel has also reported on the merits of the MIR region over the NIR. They applied MIR spectroscopy to quantify glucose and lactate in CHO cell cultures and were able to make accurate measurements over the course of 60 days without the need to recalibrate.

The amount and frequencies of light absorbed by the sample can be correlated to the type and concentration of chemical species present in the sample. The resultant absorbance spectra contain complex spectral features that require the use of chemometric methods such as partial least-squares (PLS) regression analysis to extract the analyte concentration information. These analysis methods are routinely practiced in the agricultural and petrochemical industries.

In cell cultures, NIR spectroscopy (NIRS) has been used to quantify the concentrations of glucose and maltose; glucose and glutamine; glutamine and asparagine; glucose, glutamine, ammonia, lactate, and glutamate; and glucose, sucrose, and fructose. Spectroscopic methods have recently been applied to monitor biological reactors. For yeast cell fermentations, the concentrations of ethanol; fructose, glycerol, glucose, and ethanol; and yeast cell density have been quantified. In bacterial fermentations, the concentrations of lactate, glucose, and biomass; acetate, ammonia, and the cell density; NH4OH; and exopolysaccharide, lactic acid, and lactose concentrations have been quantified. In the past 5 years, use of spectroscopic techniques has expanded substantially. Pollard and coworkers used attenuated total reflectance spectroscopy in a pilot plant to quantify sugars, phosphate, and proline for a fungal culture. Attenuated total reflectance is most useful for evaluating liquid samples with a high reflectivity. Unfortunately, many suitable crystals

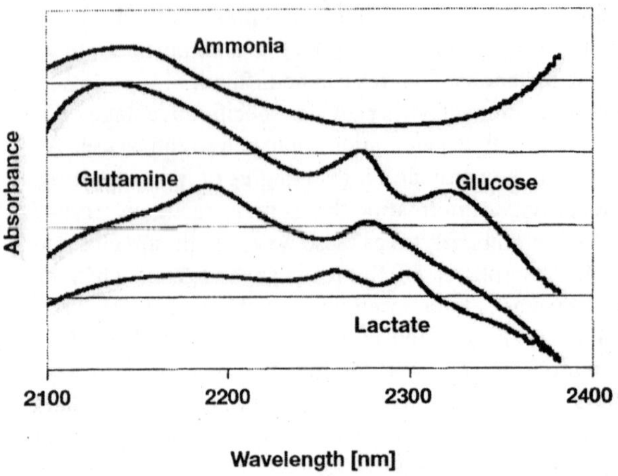

Fig. 2.7 NIR spectra of ammonia, glucose, glutamine, and lactate in a physiologically balanced salt solution. Note that each component has a distinct spectral feature that may be used for identification and quantification.

used for this method also permit strong attachment by proteins and cellular debris, thus rapidly fouling the sensing element.

Most of the aforementioned studies applied spectroscopic methods to quanti-tate the composition of yeast and bacterial cell growth media that have relatively few components compared with mammalian cell growth media. Often, the primary carbon source is present at levels of 100 mM or more in bacterial growth media. In contrast, performing such measurements on animal cell cultures is more difficult due to the greater complexity of animal cell culture media and due to the lower concentrations of many nutrients. For example, RPMI 1640 media used for animal cell culture contains 11 mM glucose, and also has 20 amino acids, in addition to vitamins, inorganic salts, and various amounts of undefined serum. There have been fewer published works on the development of NIR spectroscopic techniques for mammalian cell cultivation. Yano and Harata quantified glucose, glutamine, ammonia, and lactate in mammalian cell cultivations; similarly, McShane and Cote quantified glucose, lactate, and ammonia in cell culture media; Jung, quantified glucose using a fiber-optic measurement. Recently, as many as 19 components have been quantified simultaneously in animal cell culture media. The presence of horse serum had no apparent effect on measurement accuracy, as long as serum was present in the calibration sample sets.

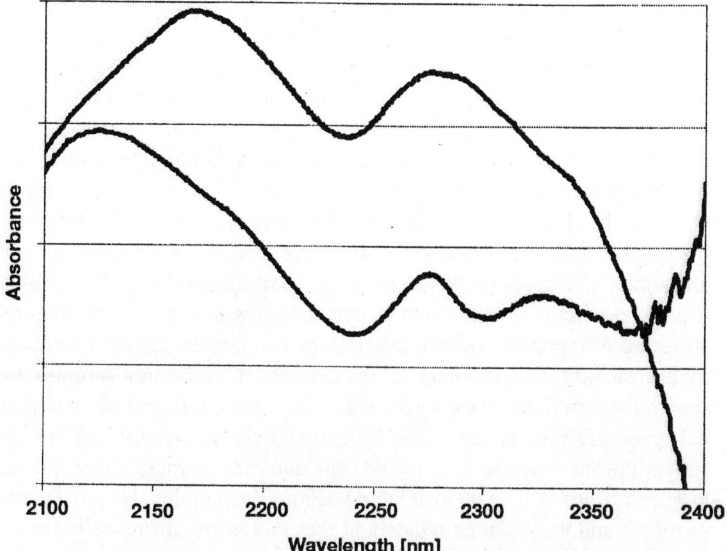

Fig. 2.8 Infrared spectra of animal cell culture media. The bottom spectrum is the commercially available DMEM; the top spectrum is DMEM + 10% horse serum.

The large number of chemical species present in mammalian cell growth media produces NIR spectra with multiple, overlapping absorbance features. PLS regression analysis is often applied to quantitatively correlate spectral features such as absorbance peak shapes and heights to the concentration of a specific analyte present in the sample. PLS regression methods require that analyte concentrations in the calibration samples must be random and uncorrelated. This requirement of uncorrelated species concentrations complicates the construction of calibration models for bioreactors.

For monitoring biological processes, the requirements on calibration samples typically are satisfied through use of 70–100 calibration samples with independent concentrations of each component. Developing samples with independent concentrations of each component, collecting spectra, and analyzing the calibration can demand many weeks of laboratory work. If any substantial modification is made to the bioprocess (pH, temperature, absence or presence of new chemical constituents), this calibration may no longer be applicable, thus requiring a lengthy recalibration process. Such difficulties have so far limited the application of NIR spectroscopy for many biological processes.

Calibration of spectroscopic techniques presents a sizeable difficulty. If a single bioreactor run is used to generate samples for calibration, unwanted correlations in the analyte concentrations will arise. If such a calibration model were applied to a second bioreactor run that has slightly different initial conditions, the spectroscopic measurements are likely not to be accurate. Samples collected from batch cultures typically contain highly correlated metabolite concentrations as a result of metabolic relations. For example, the concentrations of glucose and lactate are often anticorrelated, as are glutamine and ammonia. Calibrations based on such samples can only be reliably used to predict concentrations in new samples if a similar correlation structure was present; unusual variations can lead to highly inaccurate measurements. This problem can be avoided if calibration samples are drawn from multiple bioreactors. Fayolle and coworkers operated four calibration fermentations, while Hall and coworkers ran five calibration fermentations. The use of multiple runs does not guarantee that species concentrations will be uncorrelated because nutrient levels consistently diminish and wastes accumulate and these rates are intrinsically linked through the cell metabolism. The use of synthetic calibration samples (which are produced through addition of known amounts of each component) simplifies the removal of correlations by producing a large set of samples with uncorrelated analyte concentrations. Unfortunately, this approach requires prohibitive preparation time. This preparation time depends on the number of varying components, but can be from 4 weeks to several months.

A recognized problem for spectroscopic measurements arises when a working calibration is applied to a similar, but not identical system. Common changes that create difficulties in transferring calibrations include varying background composition, temperature alterations, and path length modifications. For bioreactors, a change in the type of culture medium could be expected to affect spectroscopic measurements. Standardization methods have been developed to reduce the time required to generate transferable calibrations for new conditions, by applying samples intermediate or common to the two measurement conditions.

Adaptive calibration schemes have also been applied. These methods often involve spiking samples with additional amounts of the analytes to be quantified. In this approach, a limited number of calibration samples are collected, typically 5–10, from bioreactors operated under normal conditions over time. These samples are each subdivided into 10–20 additional samples. To each of these is added a known amount of the analytes to be quantified. The initial large volume samples provide spectral information on uncontrolled variability including cell debris, low concentration wastes, and the breakdown of serum. The spiked material provides a greater variety of analyte concentrations and can increase the calibration range. Note that this approach can only be applied to increase the analyte levels. Rhiel applied a similar adaptive calibration procedure to quantify glucose, lactate, glutamine, and ammonia in cultures of PC-3 cells and obtained very low measurement errors.

An alternative and more rigorous approach to construct spectroscopic calibration models for bioreactor monitoring is to produce synthetic samples with each of the compounds present in the culture media, taking into account changing concentrations over the course of the cultivation. For example, at early times, glucose and glutamine concentrations would be high, whereas lactate and ammonia would be very low. This would be reflected in the synthetic samples produced. Typically, 80–100 samples are required to generate a reliable calibration, so this can be a very time-consuming procedure, particularly when taking into account the low-level amino acids and carbohydrates that typically play a minor role in cell metabolism. This approach has been successfully used to quantify 19 components simultaneously; however, calibration development time was several months.

Solution pH, temperature, and ionic strength are important experimental parameters that can affect the spectroscopic measurements. Each of these parameters can alter the position and intensity of absorbance bands in the NIR. Small band shifts can be observed due to changes in the hydrogen bonding nature of the sample caused by changes in the pH, temperature, and ionic strength. Changes in temperature by a few degrees may alter the position of the water absorbance features normally located around 1923 and 2632 nm. The pH, temperature, and ionic strength of the samples are adjusted so as to not be affected by addition of analytes and maintained at levels observed in the bioreactor. If such variations become unavoidable, digital Fourier filtering may be employed to reduce these effects. Such methods have been shown to compensate for significant alterations in absorbance features brought about by temperature changes of four degrees. The Fourier filtering step in spectral preprocessing effectively discriminates between broad baseline features and the relatively narrow analyte bands, and, hence, enhances the analyte information relative to the baseline variation.

Vaidyanathan and coworkers analyzed the sensitivity of spectroscopic calibrations to variations in analyte concentrations and to background variations. They found that simple models, such as uni- or bivariate linear regression models or PLS models using no more than four factors, performed reasonably well when subjected to variations. However, models based on weak absorbance features were vulnerable to changes in the background matrix. This study highlights the need for development of robust models, particularly without over-modeling the available calibration data. Over-modeling results from using too few calibration samples or from applying too many calibration factors. An over-modeled calibration can provide unreliable results when challenged with even small variations in the sample material.

The challenge, therefore, for applying NIRS lies in developing a calibration that provides accurate concentration measurement to which a substantial amount of confidence can be applied without the calibration being overly specific so that it can be applied to reaction schemes with small variations. A recently published study brings into question the validity of some calibration schemes for measurement of glucose from NIR spectra collected in a time-dependent manner. This group developed a "phantom" glucose data set by purposely omitting glucose in a number of samples that were collected consecutively. Within the data processing step, arbitrary and nonzero glucose values were assigned to successive phantom spectra, and multivariate calibration models generated for glucose based on PLS regression. The result was that PLS could be "tricked" into predicting the presence of glucose in samples devoid of glucose. Chance temporal correlations between assigned glucose concentrations and some uncontrolled experimental parameter are responsible for this apparent model functionality. This study presents sizeable evidence that one must take extreme care in developing calibrations that do not contain correlations between components, both quantifiable and nonquantifiable.

One of the practical constraints that may limit the application of NIRS to monitor biological processes is the limit of detection of the technique. Many of the aforementioned results provide measurement errors in the neighborhood of 0.12–1.2mM. There is no straightforward method for converting the standard error of prediction (SEP), normally provided as the comparable metric for PLS analysis, to a limit of detection (LOD). A reasonable approximation is that the LOD is three times that of the SEP (analogous to the use of three times the measurement standard deviation as a predictor of the LOD in other measurement schemes). Therefore, a 0.5 mM SEP would yield a 1.5 mM LOD. This and other information has led some researchers to believe that 1 mM is the practical limit of detection for compounds with a molar absorbtivity similar to that of glucose. Ideal cultivation conditions for hybridoma cells typical require maintaining concentrations of nutrients and wastes at concentrations as low as 1 mM each.

This evaluation of the LOD of NIRS is lacking in that the quantified SEP is highly dependent on the concentration range evaluated. For example, measurement of glucose over a concentration range of 0–30 mM yields an SEP of 0.49 mM, while measurement of glucose over a range from 0 to 1 mM yields an SEP of 0.12 mM. Clearly, the SEP does not scale directly with the concentration range, but increases at a rate less than an increase in the concentration range. Measurements of glucose and glutamine may be reliably performed at concentrations as low as 0.2 mM, the approximate level required for adequate control.

Sensing in Microbioreactors

In recent years, microscale approaches have been developed to provide information on cell cultures maintained in highly scaled down processes. Instead of performing process development on the 5–500 L scale, cultures are grown in 0.5–5 mL volume reactors. Working at such a small scale permits analysis of many cultures in parallel and in high throughput, thus providing a relatively inexpensive and fairly rapid means to evaluate a large number of process conditions simultaneously. This philosophy of high throughput has been adopted from the genomic and proteomic screening fields. Challenges in this area lay in development of suitable reactor systems, in development of analytical techniques for analyzing small sample volumes (often as little as nanoliters of material), and in analyzing scale up of such a magnitude.

Several groups have explored microbioreactors. Kostov developed 2 mL microbioreactors to evaluate cultivation conditions of *Escherichia coli* fermentations. They monitored pH, DO, and optical density using light emitting diodes, photodetectors, and an oxygen-sensing device. Their approach is simple and inexpensive and can be applied to many other types of culture systems. Their results corresponded closely to those obtained in a 1 L fermentation. Riley and coworkers have applied light emitting diodes and photodetectors as biological sensing elements to track the metabolism of lung cell cultures grown in plastic spectrophotometric cuvettes. A schematic of this reactor and sensing components. This approach is used to track the impact of environmental and biological toxins on culture metabolism, ultimately correlated to other metrics of biological toxicity. This device is small, inexpensive, and portable, thus providing a means to evaluate the safety of airborne, soluble, and particulate materials.

A significant challenge in working with microbioreactors lies in quantifying the desired products. Oftentimes, ELISAs (enzyme-linked immunosorbent assays) are used to quantify products of cell cultivations that rely on the highly specific and strong binding of a monoclonal antibody to their antigens. Typically, ELISAs require a minimum of 4–5 hr to perform and require frequent intervention by a technician or by expensive robotics. These assays can be scaled down and automated on the surface of a microchip, particularly on etched glass surfaces.

A number of groups have recently developed microchip biosensors for quantifying small compounds. Similar methods can be used to sequence DNA, although capillary immunoassays have also been developed. Quantification of as little as 2100 molecules in a single sample has been reported. Management of thermal gradients is a prime concern. These approaches employ microfabrication techniques, including photolithography and chemical etching, to generate a series of channels on the surface of a glass or silica substrate. Each channel is independently wired to permit capillary electrophoretic separation that isolates molecules based on the charge to mass ratio of each. Such fabrication techniques permit tremendous flexibility in the size of channels and on the flow rates of fluids. Each design layout may be replicated with a high degree of consistency. Sample volumes may be as small as 50 nL, which require separation times of ~30 sec. Detection methods vary, but are often based on laser-induced fluorescence. Sensitivity and the ability to detect low levels of cytokine secretion are difficulties with the current technologies. presents a photomicrograph of an etched glass surface employed for detection of cytokines produced by cell cultures. The separation between lanes is ~10 mm. This device is produced using chrome deposition on glass, followed by deposition of a photoresist that is exposed to UV light through a mask and rinsed away. The pattern is etched into the glass with a buffered oxide etch of hydrofluoric acid for 2min followed by removal of any remaining chrome. The result is a glass surface etched with the pattern of the mask.

On-chip measurements have been used to separate antibodies from unbound antigen in a competitive-binding assay in which either an antibody or antigen is labeled with a fluorescent tag. The typical operation involves adding a sample with an unknown concentration of an antigen to a solution containing an antibody reactive to this antigen and a known concentration of fluorescently labeled antigen. Any antigen present in the sample will displace an equal amount of labeled antigen from the antibody. Capillary electrophoresis separates the free antigen (both labeled and unlabelled) from the antibody-antigen complex based on the charge-to-mass ratio. Laser-induced fluorescence is used to quantify the labeled antigen whether it be bound or free, thus resulting in an electropherogram that can be related back to the concentration of antigen in the original sample. The entire procedure can be rapid (as little as 30 sec), requires minute sample volumes (around 50 nL), and has low detection limits (around 1 ng/mL). A limitation in using these techniques is the requirement of high electrical field strengths on the order of 800 V/cm. This high field strength would disrupt cellular processes and so these assays must be performed in a cell-free environment.

Electronic Noses

Electronic noses are also being applied to monitor the composition of the head space or gas exit streams from bioreactors. Bachinger and coworkers measured the off-gas composition from perfusion cultivations of a CHO cell line producing recombinant proteins. In general, this approach involves the use of multiple gas sensors that provide information on multiple analytes at one time. Often, neural networks are employed to analyze the data to search for patterns in the data. For example, the CO_2 evolution rate may be strongly correlated to the production of CH_4 or other volatile compounds. This correlation and the proportion of specific compounds produced can be specific for certain types of organisms and can be predictive of the energy state of the culture.

Cell Density

Maximizing the cell density through increases of the growth rate and cell yield is often one of the hallmarks of a successful mammalian cell culture process. Ideally, cell density measurements should be accomplished inside the bioreactor without the need to remove a cell

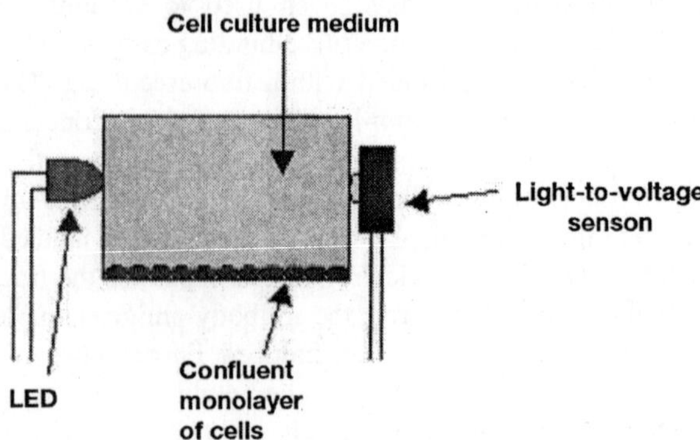

Fig. 2.9 Schematic of a microbioreactor with LED-based sensing of cell metabolism.

sample and perform microscope hematocytometer counts. Several methods have been developed in recent years to address this.

The concentration or cell density present in a bioreactor can be quantified through a number of direct or indirect means. Indirect methods assume some constant value of the cellular metabolic rate per viable cell and thus quantify the consumption of certain analytes such as glucose or oxygen, by the production of lactate, or reduction of the pH. Such approaches provide only minimal accuracy as the cellular metabolism can change over the course of a bioreactor run.

The oxygen uptake rate, glucose uptake rate, and the rate of waste generation have been applied. A major limitation of these methods is the reliance on the assumption of constant cellular metabolic rates. Time-independent metabolic models have been used to circumvent this difficulty by correlating the specific metabolic rate to the limiting substrate concentration, providing a continuous means to determine the specific uptake rate from online measurement of the limiting substrate. An advantage of this approach is that the method can be applied to determine online cell concentration in both freely suspended and immobilized cell cultures.

Direct methods of cell quantification are significantly more accurate and more robust. These can include spectroscopic measurements, turbidity measurements, and optical density measurements. Unfortunately, these measures do not provide quantification of the number or percentage of viable cells, but rather return total cell numbers. Cells do strongly absorb light at 260 nm, but, this wavelength is also strongly absorbed by proteins. IR measurements of cell density have been applied with good success. A variety of optical density probes have been evaluated for use in online monitoring of cell densities in bioreactors.

Online optical cell density probes have been used to continuously monitor the cell densities in mammalian cell bioreactors and to achieve advanced bioreactor controls. Backscattering probes were found to produce the most linear response with respect to the concentration of hybridoma cells whereas transmission probes had a lower resolution. Fouling of the probe surface due to attachment of cells is always of concern. Wu and coworkers found little fouling after 2 weeks of operation, but the presence of attached debris became apparent after 3 weeks. The impact of fouling on these measurements could be removed. A laser turbidity sensor has been used to monitor online the cell concentration in batch hybridoma cultivation with reasonably good correlation between offline cell counts and the sensor signal. Marose and coworkers recently reviewed developments in optical-density probes, in situ microscopy, optical biosensors, and fiber-optic sensors for bioprocess monitoring.

Dielectric permittivity and electrical impendence spectroscopy have been used to quantify viable biomass for mammalian, yeast, and bacteria cells with measurements being sensitive to changes in cell volume and to changes in the viable cell population.

Cell density, or biomass content, can also be quantified through spectroscopic methods. Sivakesava and coworkers compared FT-MIR spectroscopy and FT-Raman spectroscopy techniques to quantify *Saccharomyces cerevisiae* during ethanol fermentation. They report that FT-MIR was more successful because the Raman scattering of the cultures was weak, thus leading to lower useful signals.

Aseptic Sampling

As animal cell cultures require strict maintenance of sterility, all sensing and measurement must be undertaken within this constraint. Practically, this requires either that an aseptic sampling method be designed or that the sensors be sterilizable by steam of chemical methods. Long-term stability of the sensor helps to minimize the need for frequent replacement and sterilization. In general, it is preferable that the monitoring of a bioreactor be performed in situ, without the need to remove samples. Typically, such measurements can be performed with less operator intervention, do not consume culture medium or cells, and do not increase the risk for introducing contaminants. However, monitoring schemes for many desired process variables cannot be quantified in situ, thus requiring aseptic sample removal. Examples of ex situ measurements include FIA, HPLC, and GC analyses performed at a location separate from the bioreactor environment.

Aseptic sampling from liquid systems includes nonmembrane filters, dialysis membranes, spin filters, and ultrafiltration membranes. Rotating membrane samplers have been shown to provide better filtration performances than stationary samplers due primarily to a lower propensity to clog with cells or cell debris. Once collected from a bioreactor, it is preferable that the sample be transported directly to a sensing device. If allowed to incubate outside of a bioreactor for any significant time it may permit the growth of unwanted microorganisms that would skew measurements of pH, nutrients, or wastes.

An ideal situation, if a sample must be removed from the bioreactor, is to employ an autosampler, possibly consisting of a peristaltic pump, an interval timer, and a fraction collector. The autosampler repetitively removes samples of user-determined size at repetitive time intervals and injects them into the sensing device. Such an approach requires less operator intervention, thus permitting a greater degree of automatic control of the process. However, such automatic sampling often does increase the chance of introducing culture contamination and for small-scale processes the volume of material removed for measurement can be substantial compared to the total bioreactor volume. Due to these limitations, it is recommended that autosampling be applied preferentially for bioreactors for which the frequent (i.e., hourly) measurement of culture parameters can lead to a direct increase in process performance.

Acoustic filtration has also been used to retain cells within a bioreactor while still permitting frequent sampling or use of high medium perfusion rates. High media flow rates or high cell concentrations can significantly reduce the separation performance. The relative merits and limitations of other methods to retain cells and permit bioreactor sampling have been reviewed by Woodsid d coworkers.

Estimation of Rates and Metaboli Ratios from Online Measurements

Online measurements can provide a number f valuable metrics of the health and productivity of a culture.

Oxygen Uptake Rate

The oxygen uptake rate (OUR) is a good indicator of cellular activity, and even under some conditions a good indicator of the number of viable cells. Hassan and coworkers report a comparison of the viable cell counts, packed cell volume, intracellular nucleotide ratios, cell cycle analysis, online OURs, and optical density for the prediction of the end of exponential growth to optimize transfer times during scale-up of CHO cell cultures. Viable cell concentration, packed cell volume, and relative abundance of cells in S-phase were not very reliable at determining the end of exponential growth during the process; however, online determination of OUR and offline determination of intracellular nucleotide ratios (U-ratio) were very sensitive to changes in growth rate, enabling clear determination of the end of exponential growth within a short time. Optical density showed an inflection along with OUR and U-ratio but was less sensitive in determining the end of exponential growth.

Methods to quantify the OUR uptake rate and the oxygen transfer rate (OTR) begin with a mass balance on oxygen within a bioreactor. This can be written as the accumulation being equal to the rate of transfer into the culture minus that which is consumed:

$$\frac{dC}{dt} = k_L a(C^* - C) - \text{OUR} \times \text{X}$$

where $k_L a$ is the mass transfer coefficient, C is the concentration of oxygen in solution, C^* is the equilibrium solubility of oxygen (oxygen saturation), and X is the cell density (cells/L). The units on OUR are typically (g O2/106cells-time). The oxygen transfer rate is

$$\text{OTR} = k_L a\,(C^* - C)$$

The three most common methods to evaluate the delivery of oxygen to a culture involve different manipulations or boundary conditions of the mass balance. The accuracy of the OUR measurements and the required analytical devices are quite different from method to method. Note that it is advisable that control of DO concentrations be combined with pH control to avoid accumulation of CO2 and to avoid stripping of volatile compounds from the medium.

Static Method

In the static method, the consumption of oxygen is assumed to be negligible as non-respiring (or dead) cells are employed, thus simplifying the mass balance. Oxygen is removed from the bioreactor and then resupplied. This method is also called static gassing out. Specifically, a culture of cells is inoculated into the bioreactor, allowed to grow to some typical density, and then are administered either a metabolic blocking agent or a toxin such as azide. Oxygen is removed from the head space by purging with nitrogen. Aeration is turned on at the typical gas flow rate and the impeller is operated at a typical level. The increase in the oxygen concentration is followed until oxygen saturation (C^*) is reached. Thc change of the DO concentration in time is described by:

$$\ln\left(1-\frac{C}{C^*}\right) = -k_{\mathrm{L}}a \times t$$

The slope of a plot of ln(1°C/C^*) versus time gives a slope of °kLa. Care must be taken to ensure that oxygen is not consumed within the bioreactor during measurement otherwise $k_{\mathrm{L}}a$ will be underestimated. The static method provides a relatively easy and rapid means for evaluating the potential delivery of oxygen to a culture.

Dynamic Method

In the dynamic method, also called dynamic gassing out, aeration to an active culture is briefly turned off and the unsteady-state mass balance of oxygen tracked. It is critical that the oxygen concentration not drop to such a low level that the culture is negatively impacted. Often this is defined as Ccrit equal to the oxygen concentration for which the growth rate is 99% of its maximal level.

Fig. 2.13 presents a schematic description of the dynamic method. Initially, the bioreactor contains a healthy and active culture. At some time, t_0, the aeration is turned off and the concentration of oxygen decreasxes due to that consumed by the cells. A short time later (much before the oxygen level reaches C_{crit}), the aeration is turned back on. It is very important that C not approach Ccrit so that the rate of oxygen uptake is independent of the oxygen concentration. The DO concentration will steadily rise until it reaches the steady-state level (C_s). Prior to reaching the steady-state, both the DO concentration and the time are recorded twice (C_1 and t_1; C_2 and t_2). Analysis begins with the oxygen mass balance evaluated at times representing the steady-state and the unsteady-state conditions. We require expressions for both the OUR and the OTR.

The OUR can be evaluated based on the steady-state condition which yields,

$$\text{OUR} \times \text{X} = k_{\text{L}}a(C^* - C_S)$$

where C_S reflects the steady-state DO concentration. This relation can be solved for $k\text{L}a$ and substituted into the equation above to yield,

$$\frac{\text{d}c}{\text{d}t} = k_{\text{L}}a(C_S - C)$$

If $k_{\text{L}}a$ is constant over time, we can integrate the above relation, and using the boundary conditions of $C = C_1$ at $t = t_1$ and $C = C_2$ at $t = t_2$ we obtain,

$$k_{\text{L}}a = \frac{\text{In}(\frac{C_S - C_1}{C_S - C_2})}{t_2 - t_1}$$

The OUR can then be calculated from the mass transfer coefficient, the cell density, the oxygen solubility and the steady-state oxygen concentration.

The dynamic method is widely used as it can be applied without the need to significantly alter the operation of a bioreactor. Errors can arise due to slow-responding DO electrodes and due to the comparatively small oxygen uptake rate of many animal cell cultures. Measurement time should be kept significantly shorter than the doubling time of the cells so as to avoid errors introduced by changing cell densities. Singh described a computational method for the determination of oxygen uptake from the dynamic response of DO probe. For a hybridoma cell culture, the OUR for exponential growth in a standard serum-containing culture medium was 0.15mM O2/109cell-hr.

Oxygen Balance Method

A third method for quantifying oxygen transfer to a bioreactor is the oxygen balance method, which is based on an evaluation of the gas-liquid mass transfer. This approach requires measurement of the oxygen concentration of gas streams into and out of the bioreactor. A steady-state mass balance will yield the rate of oxygen transfer into the medium,

where *NA* is the rate of mass transfer into the medium, *R* is the gas constant, *V* is the volume of liquid medium, *Fg* is the volumetric gas flow rate, T is the temperature, PA is the oxygen partial pressure in the gas phase, and "in" and "out" refer to gas streams entering or leaving the bioreactor.

A primary difficulty in obtaining accurate measurements with this approach is that the partial pressure of oxygen in the streams entering or leaving the bioreactor will differ by only a small amount and thus need to be measured very accurately. The bioreactor must be operating at a steady state. Advantages of this method are that the mass transfer properties may be obtained from a single measurement and normal bioreactor operation need not be disturbed. Note that the oxygen concentration in the medium (C_s) will also necessarily be quantified so as to determine the mass transfer coefficient from:

Proper mixing and an even distribution of oxygen throughout the medium is required for an accurate measurement of C_s and hence for reliable calculations of $k_L a$. It is advisable to measure the DO concentration at multiple locations within the bioreactor to ensure an accurate value for C_s. Good mixing within the bio-reactor environment is critical.

Sulfite Oxygen

This method is based on the oxidation of sodium sulfite to sulfate by oxygen in the presence of a catalyst such as a divalent cation.This is a relatively fast reaction compared to oxygen transfer, and so is limited by the OTR. Unreacted sulfite can be removed from the bioreactor and quantified using a back titration with an iodine-starch reaction for quantification. The sulfite oxidation method yields higher kLa values than other methods and is influenced by pH and operating conditions in a difficult to characterize manner. For these reasons, this method is not highly recommended.

Measurement of OUR can be difficult due to the very low specific consumption rate (~2×10^{13} mol cell/hr), the sensitivity of the cells to variations in DO concentration, and challenges in providing variable oxygen levels without damaging the cells. This delivery issue has been resolved by using a gas-permeable membrane to provide the required amounts of oxygen delivery directly into the culture which avoids foaming of increased liquid shear stresses due to gas bubbles in solution. Unfortunately, this approach is not easily scaled up. A mass balance on oxygen is applied to quantify the OUR. For CHO cells during batch and continuous culture OUR values of 2.85×10^{-13} and 2.54×10^{-13} mol O_2/cell hr, respectively. Changes in the OUR can be

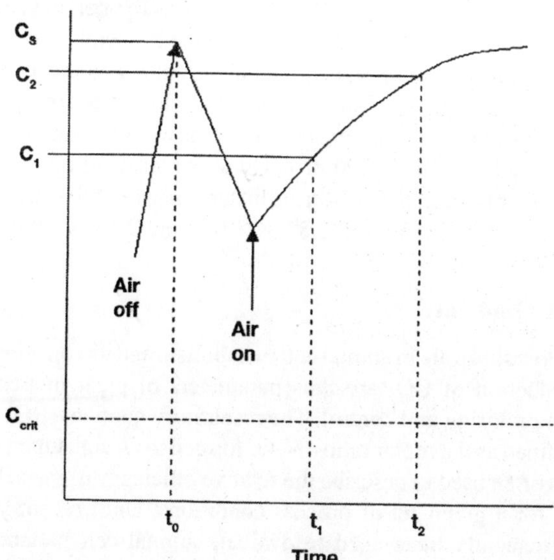

Fig. 2.10 Dynamic method for oxygen transfer measurement as described in the text.

used to track transitions in the culture behavior including the end of exponential growth.

Fed-batch cultures were implemented to study the metabolism of HEK-293 cells.Glucose, measured every 30 mm by a FIA biosensor system, was maintained at 1 mM throughout the culture using an adaptive nonlinear controller based on minimal process modeling. Maintaining a low glucose concentration, thus decreasing the rate of glycolysis, significantly reduced lactate production.

The rates of glucose and glutamine uptake as well as the lactate and ammonia production were compared to those obtained in batch mode with an initial glucose concentration of 21 mM. Basically, three phases were observed in both culture modes. The metabolic shift from the first to the second phase was characterized by a significant reduction in glucose consumption and lactate production while maximum growth rate was maintained. The specific respiration rate appeared unchanged during the first two phases, suggesting that no change occurred in the oxidative pathway capacity. In the third phase, cell growth became slower very likely due to glutamine limitation.

A system for measuring the OUR based on an oxygen balance on the liquid phase was developed. A gas-permeable membrane would provide the required quantity of oxygen into the culture, while avoiding problems of foaming or shear stress generally linked to sparging. This aeration system allowed maintenance of a known and constant kLa value through cultures up to 400 hr. OUR was measured online for a CHO cell line was determined during batch growth and continuous culture as, respcctively, equal to 2.85×10^{-13} and 2.54×10^{-13} mol O_2/cellhr.

Respiratory Quotient

Oxygen is a key substrate in animal cell metabolism and its consumption and the production of CO_2 are thus parameters of great interest for bioprocess monitoring and control. The respiratory quotient (RQ) for a culture is defined as the molar ratio of CO_2 formed to O_2 consumed. This parameter is often used to describe the relative efficiency of the cellular metabolism for a given set of process conditions. Until recent years, RQ had infrequently been used to evaluate animal cell metabolism primarily due to difficulties in measuring the CO_2 evolution rate. The oxygen uptake rate has traditionally been more readily quantified accurately.

An important complication in the determination of the CO_2 evolution rate is that standard culture media is buffered by bicarbonate, and so the CO_2 balance is affected by accumulation and therefore the RQ cannot directly be calculated from gas-phase measurements. A significant portion of the gas-phase CO_2 was found to evolve from the culture medium. The introduction or loss of CO_2 from the culture medium during preparation and sterilization must be taken into account. Modeling of liquid- and gas-phase mass balances and data filtering methods have been used to account for this CO_2 reservoir.

RQ values for mammalian cells grown under near optimal conditions are generally 1.0 ±0.05. In general, an increase in RQ has been shown to correlate with an increase in cellular productivity and so RQ could serve as a means for estimating the physiological state of the cells.

RQ values for Sf-9 insect cells have been quantified using online gas-phase O_2 measurements and IR CO_2 measurements. Linear relationships between viable cell densities and both oxygen uptake rate and carbon dioxide evolution rate were obtained in exponentially growing cultures. The extent of the increase in CER following infection and the time postinfection at which maximum CER was attained were negatively correlated with the multiplicity of infection at multiplicities below the level required to infect all the cells in a culture. The relative permittivity, carbon dioxide evolution rate, and the cell volume profiles were closely matched during the growth phase of cells when grown in a batch or fed-batch culture. The relationship became more complex when the cultures were either in stationary phase, or in the postinfection phase.

Metabolic Rates (Glucose Consumption/Lactate Production)

An important question in bioprocess modeling is how many rate equations must be specified to fully account for all relevant biological processes. The number of rate equations for which practical information can be obtained is constrained by the yield equations, which represent the balances of reducing power, energy in the form of ATP, and the various elements involved in cell metabolism. These balances are derived from a simplified picture that divides metabolism into catabolic, anabolic, respiratory, and product formation pathways. In most practical situations, a limited number of metabolic ratios are monitored and used to implement process decisions. The number of rate measurements available versus the number needed defines the state estimation problem in bioprocess control.

Stoichiometric ratios of lactate yield from glucose (Lac/Glc) and ammonium yield from glutamine (Amm/Gln) are often applied to monitor the health and productivity of a cell culture. Zeng evaluated several other stoichiometric ratios: ammonium yield from the total consumption of amino acids (NH4+/TAA), consumption of total amino acids to glutamine (TAA/Gln), essential amino acids to glutamine (EAA/Gln), glutamine to glucose (Gln/Glc), and oxygen to glucose (OUR/Glc). Several commonly employed animal cell lines (hybridoma, BHK, and CHO) demonstrated similar patterns of variation of stoichiometry. In continuous culture, Lac/Glc and Gln/Glc are primarily determined by the residual glucose concentration while TAA/Gln and EAA/Gln correlate well with the residual glutamine concentration. Ammonium formation not only is a function of glutamine concentration but also is affected by the consumption of other amino acids, particularly at low residual glutamine concentrations. NH_4^+/TAA turned out to be a more suitable parameter to describe the ammonium formation. Thus, these stoichiometric ratios could be used as a means to predict and eventually control the concentrations of nutrients that are otherwise difficult to determine online.

Metabolic flux analysis is a useful tool for unraveling relationships between metabolism and cell function. Material balancing can provide estimates of major metabolic pathway fluxes, provided all significant metabolite uptake and production rates are measured. Many serum-free media formulations contain small amounts of yeast extracts and plant or animal tissue hydrolysates consisting of meta-bolizable materials that are not easily identified or quantified. For chemostat steady states of CHO cultures grown in a steady-state chemostat culture with hydrolysate-supplemented medium, consistent flux analyses were obtained only when amino acids liberated from protein supplements were taken into account.

Since cell growth is accompanied by an enthalpy change, heat dissipation quantified by calorimetry can serve as an index for the cellular metabolic rate. Such measurements need to be performed in conjunction with the viable cell concentration measurements by an approach such as quantification of the dielectric. The ratio of the two signals gives scalar heat flux. Comparison of heat flux with glucose and glutamine fluxes indicated that the former most accurately reflected decreased metabolic activity. The set of stoichiometric coefficients in the reaction were related through the extent of reaction to overall metabolic activity. This approach can be used for medium optimization based on enrichment of amino-acids to improve cell growth while decreasing catabolic fluxes.

Application of Online Monitoring of the Cell Environment

Obtaining satisfactory performance from a cell culture requires maintenance of operating conditions at certain set points. Due to unpredictable process upsets, such as variable pumping rates, temperature fluctuations, or altered cell productivity, process parameters invariably differ from one run to the next. In some cases, such fluctuations can be predicted through previous experience with the process or through mathematical modeling of the process. Proper use of instrumentation requires a process control algorithm to determine the requisite response to process offsets.

For example, animal cells cultivated in batch systems rapidly exhaust the supply of glucose and glutamine and generate ammonia and lactate as metabolic byproducts. Several approaches have been used to extend the supply of nutrients, increase culture longevity, and maximize the final product concentration. Continuous and perfusion culture systems replace fresh medium into the reactor while removing used or spent medium, thus avoiding metabolite limitation and product inhibition. Such methods also dilute the products generated, which extends the purification steps required to obtain a concentrated product. Fortified media with abnormally high glucose and glutamine concentrations have been used in batch cultivations; however, such methods often do not substantially improve the cell density or production as the levels of ammonia and lactate increase, thus inhibiting cell productivity.

An ideal approach for maintaining extended productivity is to add nutrients to the bioreactor only as needed by the cells. Highly concentrated solutions of glucose, glutamine, and amino acids added slowly to a bioreactor have been shown to improve product concentrations by two- to fourfold over standard batch conditions. The composition of medium supplements should be based on the stoichiometric demands of the cells as determined by analysis of the spent culture medium. Such measurements should be performed rapidly so that the medium supplement matches the current cellular requirements.

Bioreactor control has been an active area of research and has attracted more attention in recent years. This is due to the new developments in related areas that can be exploited to overcome the inherent difficulties in bioreactor control. Beginning with conventional regulatory control of operating variables such as temperature, pH and DO concentration, research in bioreactor control has undergone significant changes including neural network based approaches. Rani and Rao provide a summary of recent developments in the control of batch, fed-batch, and continuous bioreactors.

Effective control requires that the characteristic times of the physical system and the control system be less than the characteristic times of the biological system. In most cases, the limiting physical characteristics are heat and oxygen mass transfer within the culture medium. These can be manipulated through changes in the power supplied to the impeller, in the impeller design, and in the bioreactor size and geometry. For most animal cell cultures, the cells maintain a doubling time on the order of 16-24h, and typically the cell response to a process alteration is much slower than that observed with cultured microbes. The response time of the sensors themselves also need to be taken into account when developing a monitoring and control scheme.

A typical monitoring and control scheme includes the following steps:

1. Observe or measure one or more process variables.
2. Compare the observed variable to a desired set point.
3. Decide what course of action is to be taken.
4. Act on the decision.

Important factors for the development of a monitoring and control scheme for a bioprocess include:

1. The rate at which a change can be effected in a process is dependent on the size of the system.
2. Sensors for many environmental variables that may impact the process are not available or may not be practical for every system.
3. Some changes can be made in only one direction. For example, one can readily increase the concentration of glucose in a bioreactor, but reducing glucose is only feasible by diluting the entire culture or by waiting for the cells to take action.
4. Many process changes that may appear to be critical for basic scientific investigation may have little impact at the production level.
5. The high degree of uniformity that may be achieved in a small laboratory scale bioreactor usually cannot be reproduced in large production vessels.

The type of control action may be either a closed loop, or feedback control, in which a correction is enacted based on observable characteristics of the process, or may follow an open loop control in which process modifications are made at predetermined times and levels. Open loop control is significantly more difficult to implement correctly for many biological processes and requires either an accurate model of the process or substantial experience in running the process. Open loop control is sometimes called preprogrammed control in that process modifications are determined prior to start of the process. For this approach to work well, the process must be highly characterized and relatively invariant from one run to the next. For example, an operator may know from experience with an individual process that over 24 hr the cell density will double, thus necessitating an increase in the impeller rotation to increase oxygen transfer. Open loop control is easy to implement and is often applied in laboratory situations; however, it is not practical for most industrial-scale processes or for processes with substantial variability. Closed loop control involves modifying one more or process parameters based on feedback obtained from sensors monitoring the process. A schematic of these processes is presented in Fig. 2.14.

Fed-Batch Systems

Mammalian cells have the ability to proliferate under a variety of nutrient environments by utilizing different combinations of nutrients, especially glucose and the amino acids. Under conditions often used in in vitro cultivation, the cells consume nutrients in great excess of what is needed for construction of biomass and desired products. They also produce large amounts of metabolic wastes with lactate, ammonia, and some nonessential amino acids such as alanine as the most prominentones. By controlling glucose and glutamine at low levels, cellular metabolism can be altered and can result in reduced glucose and glutamine consumption as well as in reduced metabolite formation.

A research group at the University of Minnesota has made tremendous strides in applying aggressive control to improve MAb production. Using a fed-batch reactor to manipulate glucose at a low level (as compared to a typical batch culture), cell metabolism was altered to a state with substantially reduced lactate production. The culture was then switched to a continuous mode and allowed to reach a steady state. At this steady state, the concentrations of cells and MAb were substantially higher than a control culture that was initiated from a batch culture without first altering cellular metabolism. The lactate and other metabolite concentrations were also substantially reduced as compared to the control culture. This newly observed steady state was achieved at the same dilution rate and feed medium as the control culture. The paths leading to the two steady states, however, were different, and so these results demonstrate steady-state multiplicity. At this new steady state, not only was glucose metabolism altered, but the metabolism of amino acids was altered as well. The amino acid metabolism in the new steady state was more balanced, and the excretion of nonessential amino acids and ammonia was substantially lower. This approach of reaching a more desirable steady state with higher concentrations of cells and product opens a new avenue for high-density- and high-productivity-cell culture.

This and similar approaches target the ideal bioreactor operation in which metabolites are added to the bioreactor only as needed by the cells as determined by an analysis of the culture medium and the

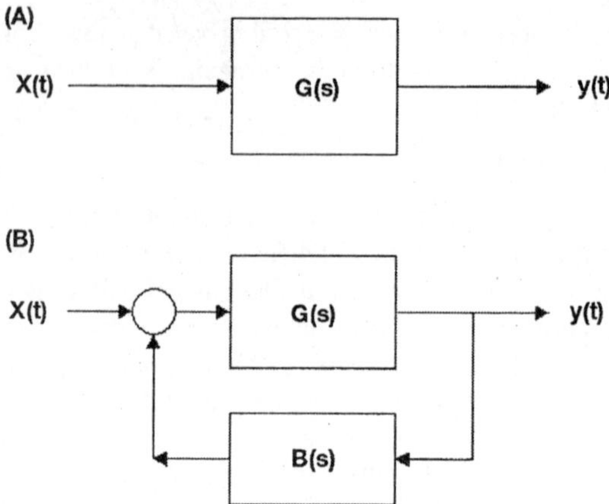

Fig. 2.11 Schematic of (A) open loop control, (B) closed loop control. Open loop control is significantly easier to implement, but requires substantial knowledge of a process. Closed loop requires frequent monitoring of the process. x(t) represents a process input, G(s) is the process, and y(t) is the process output. B(s) represents the monitoring mechanism that feeds into a comparator and decision support component.

stoichiometric demands of the cells. This is determined either through use of an online monitoring scheme, or more commonly, through offline analyses of culture medium constituents. Measurements should be performed rapidly so that the medium supplement matches the current cellular requirements. In most cases, decisions to add individual nutrients or medium supplements are done ad hoc by a bioprocess engineer, although such process decisions are likely to be automated in the near future.

The addition of highly concentrated solutions of glucose, glutamine, and amino acids can substantially improve product concentrations. Increases in product concentrations were observed by two- to fourfold over standard batch conditions. Oh and coworkers developed an interactive system for controlled feeding of glucose and glutamine to hybridoma cells so as to maximize antibody production.

A more sophisticated control scheme applies concentrated solutions of key nutrient components fed periodically to cell cultures using a simple feeding control strategy based on the integral of viable cell concentrations over time. This approach assumed constant specific nutrient consumption rates. Through effective nutritional control, both cell growth phase and culture lifetime were prolonged significantly.

These studies suggest close relationships among nutrient depletion, cell metabolism transition, and cell death.

Fed-batch cultures of HEK-293 cells were monitored using an FIA glucose measurement and controlled at a concentration of 1 mM using an adaptive nonlinear controller based on minimal process modeling. Lactate production was significantly reduced by maintaining a low glucose concentration, thus decreasing the rate of glycolysis.

A number of schemes have been applied to minimize the instantaneous concentration of inhibiting metabolites present in a culture so as to increase the yield of product in fed-batch bioreactors. Schwabe and coworkers applied a balanced supply of substrates from concentrated feed solutions based on offline estimated oxygen and glucose uptake rates. While implementation of this control scheme increased product concentration threefold, inhibition by ammonia could not be avoided. To address this difficulty, a dialysis membrane was used to remove ammonia and in which a concentrated medium was fed to the cells and inhibitory metabolites removed into a buffer solution. This improved approach led to a 10-fold increase of the product concentration compared with batch cultures.

Perfusion Systems

Continuous and perfusion culture systems replace fresh medium to the reactor and remove spent medium, thus avoiding metabolite limitation and product inhibition. Such methods also dilute products, thus extending purification steps required to obtain a concentrated product. The key parameter in perfusion culture is the rate of medium replacement (D) and this is often based on the required rate for feeding a single nutrient such as glucose or glutamine. Increases in this dilution rate can lead to an increase in the viable cell density, Xv, and to increases in the cell growth rate. Such an approach has the added advantage of improving the amount of product generated without the need for increasing the bioreactor scale, provided that both viable cell yield per perfusion rate and specific cellular productivity remain constant at higher D.

Significant reductions in cellular productivity have been observed at high dilution rates due to the high gas sparging rate needed to meet the oxygen demand. Apparently, increases in hydrodynamic shear stress imparted to the culture via intensification of gas sparging resulted in a gradual increase in specific glucose consumption and lactate production rates, while no variations were observed in glutamine-consumption rates. As a result, while glutamine was the sole limiting-nutrient under nonsparging conditions, both glutamine and glucose became limiting under sparging conditions.

Alternatively, the dilution rate does not need to be a constant, but can be manipulated based on the feedback obtained through online measurements of nutrients, wastes, cell density, or cell viability. Controlled feeding of nutrient supplements to a perfused hybridoma cell culture was used to enhance monoclonal antibody productivity. This controlled-fed perfusion approach significantly increased the volumetric antibody productivity by nearly twofold over the perfusion process, and surpassed fed-batch and batch processes by almost 10-fold. This increase in productivity can be attributed to both increased cell density as well as reduced product dilution.

Perfusion cultures of CHO cells have been performed using an acoustic filtration system to retain the cells within the bioreactor. High packed cell volumes could be maintained without additional bleeding off of cells. Perfusion of up to 50 days was performed without loss of performance of the acoustic filter. Using the protease-sensitive product rhesus thrombopoietin, cultivation in perfusion mode drastically reduced proteolysis when compared to a batch culture without addition of protease inhibitors such as leupeptin.

If high perfusion rates are to be employed, cells are often immobilized within a rigid support material in order to avoid cell washout while still permitting removal of wastes. A disadvantage of this system is that the cell density cannot be directly monitored. One group has been able to obtain high cell density of immobilized hybridomas in a bubble-column bioreactor filled with hollow glass cylinders. The parameters monitored during the cultivation were pH, temperature, DO, glucose, lactate, and monoclonal antibody. The glucose uptake rate was used to estimate the cell concentration along the time using offline measurement techniques.

Feed-Forward Control Using Mathematical Relations

Prior to designing a control strategy for a biological system, one typically constructs a simplified process model that is used to provide insight into the process dynamics. This model is based on conservation of mass and energy which may be approximated by linear, time-dependent equations with the variables that are to be controlled taken as the dependent variables. The variables that are to be adjusted so as to keep the controlled variable at its set point are the manipulated variables. For example, a mass balance on glucose in a continuous bioreactor may be as follows:

$$V\frac{\mathrm{d}C(t)}{\mathrm{d}t} = F\left[C_i(t) - C(t)\right] - \frac{FX}{Y_{X/S}}$$

where V is the volume of the bioreactor, $C(t)$ is the time-dependent concentration of glucose, $C_i(t)$ is the concentration of glucose in the feed, F is the volumetric flow rate, X is the cell density, and $Y_{X/S}$ is the yield of cells from glucose. The flow rate and feed glucose concentrations are variables which are manipulated so as to maintain the control variable, the glucose concentration in the bioreactor, at its set point.

A difficulty in applying traditional reactor control schemes developed from the chemical process industry is that even for mature bioprocesses, fundamental understanding of cell growth and metabolism remains quite limited. We are far from developing biological control models based on first principles; rather, most models remain highly empirical. The field of metabolic engineering has great promise in bridging this gap, but application of such methods directly to process control remains a challenge addressed by a small number of investigators. The problem in developing fundamental process control models is exacerbated by the highly nonlinear behavior of cell culture, the highly interactive nature of critical process variables, and the batch or fed-batch nature of most commercial processes.

Bioreactor control schemes are often based on one of two approaches. The first relies on a biological model that characterizes the cellular nutritional requirements. The control scheme attempts to match these target requirements based on the known input of cells and metabolites to the bioreactor, the expected cell growth, and the expected decrease in nutrients through metabolism and spontaneous degradation. Glacken and coworkers showed that maintenance of glucose and glutamine concentrations based on a metabolic model improved monoclonal antibody production 10-fold over batch cultivations. This approach is limited by the need for kinetic models that accurately describe cell growth and metabolism. The second control approach requires frequent measurements of the composition of the growth medium which are used to adjust the supply of metabolites added to the bioreactor to minimize the production of wastes. This approach is limited primarily by the availability of rapid measurement schemes that can quantify multiple species.

Conventional controller designs employ Laplace transforms and so require the use of linear equations. Biological processes are inherently nonlinear and have a multimodal character. Linear approximations are often applied over short time frames and for limited operating conditions. Most proposed methods fail to find the global optima for control of such processes. Even if a globally optimal control protocol can be designed for a specific process, in many instances the implementation of the control scheme presents an even more significant challenge than manual control.

Novel Strategies for Manipulating Cellular Activity

In recent years, the development of advanced systems for bioprocess monitoring and control has become an area of intensive research. Along with traditional techniques, there are several new approaches that are increasingly being applied to bioprocess operations. A detailed description of control strategies is beyond the scope of this chapter. Several examples of these novel approaches are summarized below.

1. Expert systems—Expert systems utilize an extensive database of knowl edge about a specific process and how it has performed historically. This has been used to develop process control and bioreactor state prediction based on this extensive knowledge.
2. Genetic algorithms—Roubos and coworkers applied an evolutionary program, based on a genetic algorithm to calculate optimal control policies for bioreactors. This genetic algorithm (GA) is used as a nonlinear optimi zer in combination with simulation software and constraint handling pro cedures. A GA searches parameter space by manipulating, cross-breading, and translocating "genes" which represent individual parameter settings.
3. Stochastic optimization—Stochastic optimization has been proposed as a reliable alternative to designing a control scheme. These stochastic algorithms are used to successfully solve case studies taken from the recent literature. The advantages of these alternative techniques include ease of implementation, global convergence, and good computational efficiency.
4. Ant colony algorithm—The ant colony algorithm, which mimics the coop erative search behavior of ants in real life, has been employed for the dynamic optimization of fed-batch bioreactors. The algorithm rapidly converges to optimal feed rate profiles, which maxi-

> mize the overall production of the desired product and the profits in a computationally effi cient and robust manner. The optimal profiles evolved are easy to imple ment in plant operation.

Other more traditional predictive control schemes have been applied to determine the required medium feed rates for perfusion bioreactors Adaptive software routines were developed to estimate the current and predict the future glucose uptake and lactate production of hybridoma cell cultures in hollow fiber systems. The current and future glucose uptake rates were used to select the perfusion feed rate in a designed response to deviations from the set point values. The use of the predictive controller routine decreased the glucose and lactate concentration variances up to sevenfold, and antibody yields increased by 10–43%.

The dynamic open-loop control of bioprocesses has been studied by a number of groups. Development of suitable control schemes can be performed by using a control vector parameterization concept, which makes use of second-order sensitivities to obtain exact gradients for the objective function of the underlying dynamic process model. Extensions of this method can result in efficient methodologies for solving general dynamic optimization problems, even for high levels of control discretization as typical of most bioprocesses.

Wang and Cheng claim to have developed an efficient method for simultaneously determining optimal feeding rates and operation parameters for cell cultures based on a finite dimensional optimization using the control parameterization technique. The optimal production rate obtained by the simultaneous optimization approach could be significantly improved with comparison to a simplified optimization problem.

One research group has hypothesized that inaccuracies in process optimization can be traced back to limitations in the process models applied. This group formulated a control algorithm based on cell mass and lactate production and calculated the process-model mismatch at each sampling time. These deviations were used to reoptimize the substrate concentrations throughout the length of the cultivation. The cell mass produced using dynamic optimization was compared to the cell mass produced for a nonoptimized case, and for a one-time optimization at the beginning of the batch. A single offline optimization of substrate concentration at the start of the batch increased the yield of cell mass by 27% over a nonoptimized fermentation, whereas frequent reoptimization increased yield of cell mass per batch by 44% compared to the single offline optimization. Monoclonal antibody productivities also tracked with cell densities, thus suggesting that frequent revision of control parameters can improve performance of bioreactors.

An issue that is being recognized as of tremendous concern but is only recently addressed is the role process changes have in altering not just the concentration of the desired product but also the functionality and activity of the product. Moran and coworkers developed an approach to validate ranges of control parameters for a cell culture process producing a monoclonal antibody. Specifically, the structure and functional activity of a monoclonal antibody produced at the numerical boundaries of fed-batch culture control parameters were examined with the goal of assuring that antibody produced under varying culture conditions was of consistent quality. All antibody preparations were identical to each other and to the current antibody reference standard or control. Glycosylation analysis of certain samples from the study demonstrated that the distribution of glycof orms of the antibody was not affected by the varying process control conditions of the fed-batch cultures.

Process efficiency can be strongly influenced by the cellular state that should be monitored, interpreted, and controlled. In most control systems the cellular state is not explicitly considered, rather control schemes are based on the implicit assumptions that the growth environment will directly map to the process efficiency. This limitation is well realized and explicit monitoring and control of cellular physiology are considered to be among the most challenging tasks of modern bioprocess engineering. For example, the onset of widespread apoptotic (gene-directed) cell death may be difficult to predict but may have a sizeable impact on culture productivity. The physiological state of a culture can be composed of several process variables including specific metabolic rates, metabolic rate ratios, and others. The real-time monitoring of many of these is possible using commercial sensors and has been described above. This approach reduces the goal of a control scheme to maintaining the physiological state of the cell as close as possible to an "ideal" trajectory, providing maximum efficiency for the culture as a whole. For a detailed description of this approach see Konstantinov.

3

CULTURAL MEDIA

The development of culture media for mammalian cells has been studied for more than 50 years. The first attempts at culturing animal cells in vitro made use of biological fluids, such as serum and other blood or tissue extracts. This was followed by the attempt to culture animal cells in defined media through the analysis of the contents of biological fluids. Another approach, developed by Eagle, consisted of finding the minimum ingredients that were essential for growth, which led to the development of Eagle's minimal essential medium (EMEM). This medium consisted of 13 amino acids, 8 vitamins, 6 ionic species, and dialyzed serum to provide the necessary undefined components required for growth.

As new cell lines became available in the scientific community, new formulations were developed. Many of these cell lines could be cultured in Eagle's MEM, while others required more complex formulations. These included Dulbecco's modification of Eagle's medium (DMEM), F12 medium, and Roswell Park Memorial Institute (RPMI) medium. As progress was made in the understanding of cell metabolism and growth factor requirements, various serum-free formulations were also developed.

Presently, there are many formulations available for the culture of animal cells. The decision of which formulation to use is dependent on the purpose of the culture. For the production of viruses or other nonspecific molecular studies, basic formulations such as serum-supplemented MEM are often used. However, for other studies where undefined components can affect the results, or in large-scale production systems where productivity is an issue, serum-free formulations are relied upon.

The supplementation of culture media with undefined components, such as serum, has many inherent disadvantages. This has fueled the demand for better serum-free media in both the research and the industrial communities. New formulations are being developed all the time, and there are presently many available from various commercial suppliers. However, the performance of many of these formulations remains poor as many of them contain undefined components that can affect their quality and consistency.

This chapter focuses on the composition of classic animal culture media, with emphasis on the development of serum-free formulations.

Culture Media

Basal Media

The components of these media formulations include a complex mixture of carbohydrate, amino acids, salts, vitamins, hormones, and growth factors.

The media originally used for the growth of animal cells were based entirely on biological fluids such as plasma and embryonic extracts. However, the use of chemically undefined media suffers from the disadvantages of batch variation and vulnerability to contamination. This led to a need for chemically defined culture media, which were originally based on the analysis of plasma, and this led to complex formulations such as Medium 199, which contains over 60 synthetic ingredients.

An alternative approach in media design is to reduce the number of components to the minimum shown to be essential for cell growth. This strategy led to Eagle's basal medium (BME), which was designed for the optimal growth of mouse-L cells and HeLa cells. It should be noted that there are several versions of BME. This was later modified and improved as EMEM, which has also found wide application for the growth of a variety of cell lines.

Although there are numerous media formulations now available for cell growth, the following list indicates those that have found greatest application for the cell technologist. The choice of a particular medium for a cell type has to be empirical but those listed are the ones that would normally be a starting point for attempting the growth of any new cell line.

Simple Basal Media

Table 3.1 shows the composition of several commonly used basal culture media developed for mammalian cell cultures.

- *BME* was originally designed for the growth of mouse-L and HeLa cells and has been used subsequently for the growth of a wide range of cell lines including human diploid cells. The medium needs to be changed at least every other day to support continued cell growth. It should be noted that there are several versions of BME.
- *EMEM* was developed as an improvement of BME for the optimal growth of a wide variety of cell lines including clonal cultures. EMEM has a higher concentration of amino acids than BME. It contains a balanced salt solution (Earle's); alternatively Hanks' salt solution can be used. The medium can support cell growth for several days.
- *Glasgow's modification of Eagle's medium (GMEM)* is a modification of BME and contains 2× the concentrations of the amino acids and vitamins with extra glucose and bicarbonate. This medium was originally developed for the growth of baby hamster kidney 21 (BHK21)/C13 cells and is usually supplemented with 10% tryptose phosphate broth as well as serum.
- *Joklik's modified Eagle's medium* is a general-purpose growth medium for suspension cultures.
- *Alpha modified Eagle's medium* was originally used to study ribosomes in mouse-hamster hybrid cells. The medium, which contains both essential and nonessential amino acids, can be used as a general-purpose growth medium.
- *DMEM* has a high nutrient concentration and includes 4× the BME concentration of amino acids and vitamins as well as additional nonessential amino acids and trace elements. The original formulation contained 5.6 mM glucose but 25 mM has been proved to be optimal for the growth of several cell types. It was first reported for the culture of embryonic mouse cells but has since found a wide application in the culture of various

TABLE 3.1 SIMPLE BASAL MEDIA FORMULATIONS USED IN MAMMALIAN CELL CULTURE

Components	Molarity (m M)			
	BME	GMEM	DMEM	RPMI (1640)
Inorganic salts				
Calcium chloride (CaCl2)	1.8	4.5	1.8	—
Calcium nitrate ($Ca(NO_3)_2$ 4H2O)				0.424
Cupric sulfate ($CuSO_4 \cdot 5H_2O$)				
Ferric nitrate ($Fe(NO_3) \cdot 3-9H_2O$)		0.25	0.25	
Ferrous sulfate ($FeSO_4 \cdot 7H_2O$)				
Potassium chloride (KCl)	5.3	37.33	5.3	5.3
Potassium nitrate (KNO_3)				
Magnesium chloride ($MgCl_2$ anhydrous)	0.5	11.26		
Magnesium sulfate ($MgSO_4$)	0.81	0.813	0.407	
Magnesium sulfate ($MgSO_4$ anhydrous)		11.31		
Sodium chloride (NaCl)	117	110	110.34	103.44
Sodium bicarbonate ($NaHCO_3$)	26.2	4.17	44.1	23.8
Sodium phosphate ($Na_2HPO_4 \cdot H_2O$)				5.63
Sodium phosphate ($Na_2HPO_4 \cdot$H2O)	1.01	7.43	0.906	
Energy metabolism				
D-Glucose	5.55	3.88	25	11.1
Sucrose		78		
D-Fructose		2.22		
Fumaric acid		0.474		
α-Ketoglutaric acid		2.53		
Malic acid		5		
Succinic acid		0.508		
Amino acids				
β-Alanine		2.2		
L-Alanine		2.5		
L-Arginine-HCl	0.1	3.3	0.398	1.1
L-Asparagine	2.65	0.379		
L-Aspartic acid	2.63	0.15		
L-Cystine-2HCl	5.0×10^{-2}	9.2×10^{-2}	0.2	0.206
L-Glutamic acid		4.08	0.136	
L-Glutamine		4.11	4	2.05
Glycine		8.67	0.399	0.133
L-Histidine	5.2×10^{-2}	16.2	0.2	9.7×10^{-2}
L-Hydroxyproline				0.153
L-Isoleucine	0.198	0.382	0.802	0.382
L-Leucine	0.198	0.573	0.802	0.382

Table 3.1 Simple Basal Media Formulations used in Mammalian Cell Culture (*Continued*)

Components	Molarity (m M)			
	BME	GMEM	DMEM	RPMI (1640)
L-Lysine-HCl	0.2	3.42	0.798	0.219
L-Methionine	0.05	0.336	0.201	0.101
L-Phenylalanine	0.1	0.909	0.4	9.1×10^{-2}
L-Proline	3.04	0.174		
L-Serine	10.5	0.4	0.286	
L-Threonine	0.202	1.47	0.078	0.168
L-Tryptophan	2.0×10^{-2}	0.49	0.078	2.5×10^{-2}
L-Tyrosine-2Na·$2H_2O$	10.0×10^{-2}	0.276	0.398	0.11
L-Valine	0.2	0.855	0.803	0.171
Vitamins				
Biotin	4.0×10^{-3}	4.1×10^{-5}	8.0×10^{-3}	
D-Calcium pantothenate	2.0×10^{-3}	4.2×10^{-5}	8.3×10^{-3}	5.0×10^{-4}
Choline chloride	7.1×10^{-3}	1.4×10^{-3}	2.9×10^{-2}	2.1×10^{-2}
Folic acid	2.2×10^{-3}	4.5×10^{-5}	9.1×10^{-3}	2.2×10^{-3}
i-Inositol	1.1×10^{-2}	1.0×10^{-4}	4.0×10^{-2}	1.9×10^{-1}
Nicotinic acid (Niacin)		1.0×10^{-4}		
Nicotinamide	8.1×10^{-3}	3.3×10^{-2}	8.1×10^{-3}	
p-Aminobenzoic acid		1.0×10^{-4}	7.2×10^{-3}	
Pyridoxal-HCl	4.9×10^{-3}	9.7×10^{-5}	2.0×10^{-2}	4.8×10^{-3}
Pyridoxine hydrochloride				
Riboflavin	2.7×10^{-4}	5.3×10^{-5}	1.1×10^{-3}	5.0×10^{-4}
Thiamine-HCl	2.9×10^{-3}	5.9×10^{-5}	1.2×10^{-2}	5.3×10^{-5}
Vitamin B_{12}				3.7×10^{-6}
Lipids and derivatives				
i-Inositol	1.1×10^{-2}	1.0×10^{-4}	4.0×10^{-2}	1.9×10^{-1}
Glutathione (reduced)	3.2×10^{-3}			
Indicators				
Phenol red	2.5×10^{-2}		3.5×10^{-2}	1.3×10^{-2}

cells, including primary mouse and chicken cells, and in virus production

- *RPMI 1629 or McCoy's 5A medium* was based on BME with the amino acid and vitamin mixture from Medium 199. The medium was originally formulated and later modified by Hsu and Kellogg and Iwakata and Grace. The latter modification differs in the use of D-alanine rather than l-alanine. In another modification, Park and Terasaki added a further 25 mg/L phenol red in order to observe pH changes more readily. McCoy's medium has been used as a standard medium for cloning cells.
- *RPMI 1630* was developed in 1968 by Moore and Kitamura
- *RPMI 1640* was developed for the long-term culture of peripheral blood lymphocytes. It was developed as a modification of McCoy's

5A medium. It has now been recognized as a general-purpose medium particularly for lymphocyte and hybridoma cultures. There are two commonly used modifications of RPMI 1640 that can enhance growth promotion or lymphocyte stimulation.

- *Searle's modification of RPMI 1640* includes Na_2HPO_4 (399mg/L) and $NaH_2PO_4 \cdot 2H_2O$ (440mg/L).
- *Dutch modification of RPMI 1640* includes NaCl (6400 mg/L), $NaHCO_3$ (1000 mg/L), and Hepes (4770 mg/L).

Other Simple Media

- *Fischer's medium* was originally used to support the growth of mouse leukemia cells in suspension culture.
- *Leibovitz L-15 medium* was designed for use without the need for an enriched CO_2 atmosphere. It has been used to support the growth of human diploid fibroblasts.
- *Trowell 's T-8 medium.*
- *Williams' medium E* was originally formulated for the long-term culture of adult rat liver epithelial cells.

Complex Media Intended for Use as Serum-Free Formulations Examples of these media compositions are listed in Table 3.2:

- *Biggers' Medium* was originally designed to support isolated hepatocytes in culture. Additional amino acids and vitamins led to the Fitton-Jackson modification, which has been used to support cultures of embryonic bone.
- *Connaught Medical Research Laboratories (CMRL) 1066 medium* is a complex medium designed to support cell growth without the addition of serum. The formulation represents an extensive modification of Medium 199.
- *eRDF* is based on a mixture of RPMI 1640:DMEM:F12 (2:1:1). This has been found to be effective with a serum-free formulation for the support of hybridomas. The enriched form of RDF contains 3 × the original concentration of amino acids and 2 × the concentration of glucose. Such a formulation with the addition of ITES (insulin, transferrin, ethanolamine, and selenite) can be more effective for myeloma or hybridoma cell growth than a serum-based medium.
- *Ham's F10 medium* was originally developed for the clonal growth of differentiated cells from chicken embryos. Modifications from

the original formulation have included a substitution of MgCl2 for $MgSO_4 \cdot 7H_2O$ and a 10-fold increase in the concentration of phenol red.

- *Ham's F12 medium* has a complex composition including various trace elements and was originally designed for cloning diploid hamster ovary cells. It was originally designed as a serum-free formulation but now is commonly used with a serum supplement to support the growth of a variety of normal and transformed cells. F12 has been combined with DMEM (1:1) as the basis for development of serum-free formulations. The idea was to combine the richness of F12 with the high nutrient concentrations of DMEM.
- *Ham's F14 medium* was originally developed to allow the study of sensory neurons.
- *Iscove's modified Dulbecco's medium (IMDM)* is a modification of DMEM containing additional amino acids and vitamins, selenium, sodium pyruvate, Hepes, and potassium nitrate instead of ferric nitrate. This medium has proved useful as a basis for serum-free formulations and has been widely used for the growth of lymphocytes and hybridomas.
- *MCDB 104* was developed from Ham's F12 as a medium for the clonal growth of human diploid fibroblasts.
- *MCDB 110* was developed from MCDB 104 for the growth of human diploid fibroblasts. The modifications included the addition of fatty acids, phospholipids, growth factors, hormones, and other supplements to simulate the action of serum.
- *MCDB 153* was developed as a basal medium for the serum-free growth of human epidermal keratinocytes and other epithelial cell types.
- *Medium 199* was formulated in an attempt to produce growth medium free from added protein, although for the growth of some cell lines serum is still added. This in an extremely complex medium based on Earle's salt solution and containing an extensive range of amino acids, vitamins, nucleic acid derivatives, growth factors and lipids.
- *NCTC 135 medium* was originally formulated for the serum-free growth of mammalian cells over long periods. The medium is a modification of the original NCTC 109 mediumv.
- *Waymouth's medium MB 752/1* was originally designed for the growth of L929 cells under serum-free conditions. However, it

TABLE 3.2 COMPLEX BASAL FORMULATIONS USED FOR SERUM-FREE MEDIA

Components	Molarity (mM)							
	CMRL (1066)	DMEM/ F12	F12 (Ham's)	IMDM	MCDB 153	MCDB 131	Media 199	Waymouth's MB 752/1
Inorganic salts								
Calcium chloride ($CaCl_2$)	1.8	1.05	0.299	1.49	0.03	1.6	1.8	0.816
Calcium nitrate ($Ca(NO_3)_2 \cdot 4H_2O$)		1.20×10^{-7}						
Cupric sulfate ($CuSO_4$-$5H_2O$)		7.80×10^{-9}	1.0×10^{-5}		1.1×10^{-5}	5.0×10^{6}		
Ferric nitrate ($Fe(NO_3)_3 \cdot 9H_2O$)		1.2×10^{-4}					1.7×10^{-3}	
Ferrous sulfate ($FeSO_4$-$7H_2O$)		1.5×10^{-3}	0.003		5.0×10^{-3}	1.0×10^{-3}		
Potassium chloride (KCl)	5.3	4.16	2.98	4.44	1.5	3.97	5.33	2
Potassium nitrate (KNO_3)				7.5×10^{-4}				
Magnesium chloride ($MgCl_2$)					0.6			1.18
Magnesium chloride ($MgCl_2$, anhydrous)	0.6							
Magnesium sulfate ($MgSO_4$)	0.814	0.407				10.02	0.397	0.813
Magnesium sulfate ($MgSO_4$, anhydrous)				0.814				
Mangenous sulfate ($MnSO_4$-H_2O)					1.0×10^{-6}	1.0×10^{-6}		
Ammonium molybdate ($(NH_4)6Mn_7O_{24} \cdot 4H_2O$)					1.0×10^{-6}	3.0×10^{-6}		
Nickelous chloride ($NiCl_2 \cdot 6H_2O$)					5.0×10^{-7}	3.0×10^{-7}		
Sodium chloride (NaCl)	116	120.61	131	77.59	120	110.86	81.89	1.0×10^{-4}
Sodium bicarbonate ($NaHCO_3$)	26.2	29	14	36	14	14	14.9	26.7
Sodium meta silicate (Na_2SiO_3-$9H_2O$)					5.0×10^{-4}	1.0×10^{-2}		
Sodium phosphate ($Na_2HPO_4 \cdot H_2O$)		0.5	1		2			2.11
Sodium phosphate ($NaH_2PO_4 \cdot H_2O$)	1.01	0.453		0.906		0.5	1.01	0.46
Stannous chloride ($SnCl_2$)					5.0×10^{-7}			
Sodium selenite ($Na_2SeO_3 \cdot 5H_2O$)				6.5×10^{-5}				
Selenious acid (H_2SeO_3)					3.0×10^{-5}	3.0×10^{-5}		
Ammonium metavanadate ($NaVO_3$)					5.0×10^{-6}	5.0×10^{-6}		

Zinc sulfate($ZnSO_4$-$7H_2O$)		1.5×10^{-3}	3.0×10^{-3}		5.0×10^{-4}	1.0×10^{-6}		
Energy metabolism								
D-Glucose	3.33	17.51	10	25	6	5.56	5.55	27.2
Sodium acetate-$3H_2O$	0.61				3.7		0.61	
Sodium Pyruvate		0.5	1	1	0.5	1		
Nucleic acid derivatives								
Adenine sulfate					0.18		5.4×10^{-2}	
ATP							1.7×10^{-3}	
AMP							5.0×10^{-4}	
Adenine						1.0×10^{-3}		
2′-Deoxyadenosine	4.0×10^{-2}							
5-Methyl-deoxycytidine	4.0×10^{-4}							
2′-Deoxycytidine	4.4×10^{-2}							
2′-Deoxyguanosine	3.8×10^{-2}							
Diphosphopyridine nucleotide (NAD)	1.1×10^{-2}							
Flavin adenine dinucleotide (FAD)	1.3×10^{-3}							
Hypoxanthine (Na)		1.5×10^{-2}	3.0×10^{-2}				2.9×10^{-3}	0.184
Guanine hydrochloride							1.6×10^{-3}	
Ribose							3.3×10^{-3}	
2-deoxy-D-ribose							3.7×10^{-3}	
Thymidine	4.1×10^{-2}	1.5×10^{-3}	3.0×10^{-3}		3.0×10^{-3}	1.0×10^{-4}		
Thymine							2.4×10^{-3}	
Triphosphopyridine nucleotide (NADP)	1.3×10^{-3}							
Uracil							2.7×10^{-3}	
UTP	2.0×10^{-3}							
Xanthine (Na)							2.2×10^{-3}	
Amino acids								
L-Alanine	0.281	5.0×10^{-2}	0.1	0.281	0.28	0.03	0.28	
L-Arginine-HCl	0.33	0.7	1	0.398	1	0.3	0.33	0.36
L-Asparagine		5.0×10^{-2}	0.1	0.189	0.1	0.1		
L-Aspartic acid	0.23	5.0×10^{-2}	0.1	0.23	0.03	0.1	0.23	0.451
L-Cystine-2HCl	0.1083			0.381			0.108	

Table 3.2 Complex Basal Formulations Used for Serum-Free Media (*Continued*)

	Molarity (mM)							
Components	**CMRL (1066)**	**DMEM/ F12**	**F12 (Ham's)**	**IMDM**	**MCDB 153**	**MCDB 131**	**Media 199**	**Waymouth's MB 752/1**
L-Cysteine		0.1						0.504
L-Cysteine-HCl-H_2O	1.48	0.1	0.2		0.24	0.2	6.0×10^{-4}	6.3×10^{-2}
L-Glutamic acid	0.51	0.05	0.1	0.51	0.1	0.03	0.454	1.02
L-Glutamine		2.5	1	4	6		0.685	2.4
Glycine	0.667	0.25	0.1	0.399	0.1	0.03	0.667	0.667
L-Histidine hydrochloride H_2O	9.5×10^{-2}	0.15	10.0×10^{-2}	0.2		0.2	0.104	0.826
L-Histidine					0.08			
L-Hydroxyproline	7.6×10^{-2}						0.763	
L-Isoleucine	0.153	0.416	3.0×10^{-2}	0.802	1.5×10^{-2}	0.504	0.305	0.191
L-Leucine	0.458	0.451	0.1	0.802	0.5	1	0.458	0.382
L-Lysine-HCl	0.383	0.499	0.199	0.798	0.1	0.995	0.383	1.31
L-Methionine	0.101	0.116	3.0×10^{-2}	0.201	3.0×10^{-2}	0.1	0.1	0.336
L-Phenylalanine	0.152	0.215	3.0×10^{-2}	0.4	3.0×10^{-2}	0.2	0.152	0.303
L-Proline	0.348	0.15	0.3	0.348	0.3	0.1	0.348	0.435
L-Serine	0.238	0.25	0.1	0.4	0.6	0.305	0.238	
L-Threonine	0.252	0.449	0.1	0.078	0.1	0.101	0.252	0.63
L-Tryptophan	0.049	0.0442	1.0×10^{-2}	0.078	1.5×10^{-2}	2.0×10^{-2}	0.049	0.196
L-Tyrosine-2Na • $2H_2O$	0.26	0.214	3.0×10^{-2}	0.462	1.5×10^{-2}	0.1	0.221	0.221
L-Valine	0.214	0.452	0.1	0.803	0.3	0.103	0.214	0.566
Vitamins								
Ascorbic acid	0.284						2.0×10^{-4}	9.9×10^{-2}
α-tocopherol phosphate							1.4×10^{-5}	
Biotin	4.1×10^{-5}	1.4×10^{-5}	3.0×10^{-5}	5.3×10^{-5}	6.0×10^{-5}	3.0×10^{-5}	4.1×10^{-5}	8.2×10^{-5}
Calciferol (vitamin D_2)							2.5×10^{-4}	
D-Calcium pantothenate	2.1×10^{-5}	4.6×10^{-3}	1.0×10^{-3}	8.3×10^{-3}		2.5×10^{-2}	2.1×10^{-5}	2.0×10^{-3}
Choline chloride	3.5×10^{-3}	6.4×10^{-2}	10.0×10^{-2}	2.9×10^{-2}	0.1	0.1	3.5×10^{-3}	1.79
Folicacid	2.3×10^{-5}	6.0×10^{-3}	2.9×10^{-3}	9.1×10^{-3}	1.8×10^{-3}	1.0×10^{-3}	2.0×10^{-4}	1.1×10^{-3}
i-Inositol	2.0×10^{-4}	7.0×10^{-2}	0.1	4.0×10^{-2}	0.1	4.0×10^{-2}	2.8×10^{-4}	5.5×10^{-3}

Menadione (vitamin K_3)							5.8×10^{-4}	
Nicotinic acid (Niacin)	2.0×10^{-4}						2.0×10^{-4}	
Nicotinamide	2.0×10^{-3}	1.7×10^{-2}	3.0×10^{-4}	3.3×10^{-2}		5.0×10^{-2}	2.0×10^{-4}	8.1×10^{-3}
p-Aminobenzoic acid	3.0×10^{-4}						3.0×10^{-4}	
Pantothenate					1.0×10^{-3}			
Pyridoxal-HCl	1.0×10^{-4}			2.0×10^{-2}			1.0×10^{-4}	
Pyridoxine hydrochloride	1.2×10^{-4}	9.7×10^{-3}	3.0×10^{-4}		3.0×10^{-4}	1.0×10^{-2}	1.0×10^{-4}	4.8×10^{-3}
Riboflavin	2.7×10^{-5}	6.0×10^{-4}	1.0×10^{-4}	1.1×10^{-3}	1.0×10^{-4}	1.0×10^{-5}	2.7×10^{-5}	2.6×10^{-3}
Thiamine-HCl	3.0×10^{-5}	6.4×10^{-3}	1.0×10^{-3}	1.2×10^{-2}	1.0×10^{-3}	1.0×10^{-2}	3.0×10^{-5}	3.0×10^{-2}
Vitamin B_{12}		5.0×10^{-4}	1.0×10^{-3}	9.6×10^{-6}	3.0×10^{-4}	1.0×10^{-5}	3.1×10^{-4}	1.0×10^{-4}
Vitamin A (acetate)							3.1×10^{-4}	
Lipids and derivatives								
Cholesterol	5.2×10^{-4}						5.2×10^{-4}	
i-Inositol	2.0×10^{-4}							
Linoleic acid		1.5×10^{-4}	3.0×10^{-4}					
Lipoic acid		5.1×10^{-4}	9.7×10^{-4}		1.0×10^{-3}	1.0×10^{-5}		
Ethanolamine					0.1			
Prostaglandin E_1					2.5×10^{-5}			
Other compounds								
EGF(ng/ml)					25			
Putrescine-2HCl		5.0×10^{-4}	1.0×10^{-3}		1.0×10^{-3}	1.0×10^{-6}		
Glutathione(reduced)	3.3×10^{-2}						1.0×10^{-4}	4.9×10^{-2}
Hydrocortisone					1.4×10^{-4}			
Sodium glucuronate-H_2O	1.8×10^{-2}							
Co-carboxylase	2.1×10^{-3}							
Coenzyme A	3.3×10^{-3}							
Tween 80 (mg/L)	5.0						20	
Insulin (μg/ml)					5.0			
Dialyzed FBS (μg P/ml)					1.0			
Buffers/indicators								
Hepes				25	28			
Phenol red	5.0×10^{-2}	2.0×10^{-2}	3.0×10^{-3}	3.5×10^{-2}	3.3×10^{-2}	3.1×10^{-2}	5.0×10^{-2}	2.5×10^{-2}

has been used for the growth of many cell types, usually with serum supplementation.

Standard media formulations such as those outlined previously can be purchased from commercial suppliers (e.g., Sigma, Life Technologies, etc.), either as a liquid or as a powder. Liquid forms can come as 1 × or 10 × concentrates, the latter requiring dilution with sterile deionized distilled water. Alternatively, powdered media may be preferred, especially if large quantities are needed. In this case, the powder should be dissolved in distilled water and sterilized by filtration through a 0.22 μm filter under positive pressure. Sterilized supplements of sodium bicarbonate, antibiotics, and glutamine are often added after sterilization of the bulk of the medium. This is essential if the medium is autoclaved, as these supplements are unstable.

Water for Media Preparation

The quality of water for mammalian cell culture is critical in any process. Due to the delicacy of mammalian cells the presence of trace minerals and/or contaminants will drastically affect the performance of the culture medium, especially in serum-free cultures. Due to this, specialized water purification systems are used for water in media preparation.

The overall process of water purification involves three or four-stages. These are reverse osmosis (or distillation), charcoal filtration, deionization, and micropore filtration. The first stage is typically reverse osmosis but distillation can also be used. This provides a relatively pure water supply to begin the process. Charcoal filtration will remove the majority of organic and inorganic impurities from the water, followed by deionization to remove any remainder of trace metals or ions. The final product is then passed through a micropore filter to remove any possible microbial contamination. In most systems the purified water will be continuously recycled throughout the filtering system, resulting in increasingly pure water.

The purity of the water is measured by its ability to conduct an electrical charge, in this case its resistance in MΩ cm. Most purification systems will come with a resistance meter display. The ISO 3696 standard for type-I water resistance is designated at ≥20MΩ cm at 25°C.

To ensure consistency of the water, the cartridges should be replaced on a regular basis (specified by the manufacturer). Furthermore, any tubing or storage containers used with the system should be cleaned regularly to prevent the introduction of contaminants.

For the purposes of making culture media the water should be taken directly from the purification system during media preparation. Alternatively, the water can be stored if it is sterilized before hand. This can be accomplished by autoclaving or by micropore filtration (20 μm pore size). For autoclaving the water should be placed in a suitable container, such as a pyrex (borosilicate) bottle.

Antibiotics

Antibiotics are often included in media for short-term cultures. The use of antibiotics for routine subculture or in stock cultures should be avoided as this may mask a low-level contamination that may cause problems at a later date. Furthermore, extensive use of antibiotics may cause the selective retention of antibiotic-resistant contaminants in the laboratory. When antibiotics are used, the following cocktail may be recommended:

- Penicillin (100 IU/mL) to inhibit the growth of gram-positive bacteria
- Streptomycin (50 μg/mL) to inhibit the growth of gram-negative bacteria
- Amphotericin-B (25 μg/mL) as an antifungal agent.

Nutrients

Glucose

The primary carbohydrate normally included in mammalian culture media is glucose, routinely supplemented from 5 to 25 mM. The majority of glucose is metabolized via glycolysis to form pyruvate, which is then reduced to lactate. This eventually leads to a buildup of lactic acid in the medium. Studies of metabolic pathways have revealed that only a small portion of glucose (~20–30%) enters the other pathways including the tricarboxylic acid cycle and the pentose phosphate shunt.

In addition to glucose evidence has shown that mammalian cells also utilize glutamine as their major energy source. Glutamine is often included at high concentrations relative to the other amino acids (2-4 mM), and has been shown to be a high requirement of many cell lines. In most cultures, glutamine and glucose are utilized particularly rapidly and can cause cell growth limitations even before their complete exhaustion. This trend is more pronounced in serum-free or low-protein media, and higher utilization rates have been observed.

Amino Acids

The nutrient requirement of cultured cells includes the essential amino acids (i.e., those not normally synthesized in mammals in vivo) along with others depending on the specific requirements of the individual cell line. The nonessential amino acids are usually included because some cell lines may not be able to produce their own, which would lead to a limitation in their growth potential. The effect of an amino acid limitation in the medium will be a reduction in the growth rate and/or the maximum cell density. Other nonessential amino acids can be produced by the cell but not in sufficient quantities to maintain maximal growth. Some amino acids are not stable in the medium (e.g., glutamine), thus requiring supplementation to maintain a suitable level.

Amino acids are often added as defined components into the medium. There are, however, undefined sources of amino acids, such as serum, tryptose phosphate broth or lactalbumin, or various plant and animal cell hydrolysates.

Glutamine is routinely used in culture media as it has been shown to be a major energy source for mammalian cells. However, glutamine can lead to a buildup of ammonia in the medium, which can result in growth inhibition of some ammonia-sensitive cell lines. In this case glutamate, or glutamine dipeptides (ala-gln and gly-gln), have been shown to be successful substitutes and will lower the accumulation of ammonia in the medium.

Branched amino acids are consumed particularly rapidly by a number of cell lines including Madin-Darby canine kidney (MDCK) cells, human fibroblasts, mouse myeloma cells, and BHK cells. It has been found that branched amino acids are more extensively oxidized by hybridoma cells when the specific glutamine utilization rate is low. This group also observed that the serine consumption was higher when the glutamine consumption was reduced.

Some cell lines can be more sensitive to amino acid limitations than others and can be important for feeding strategies in larger-scale systems. For example, under low glutamine concentrations, BHK cells will increase their consumption of other amino acids, especially the essential amino acids leucine, isoleucine, and valine and the nonessential amino acids serine and glutamate.

Salts

The media salt concentration is isotonic to prevent osmotic imbalances. The osmol-ality of standard growth media is approximately 300 mOsm/kg and is optimal for most cell lines. Care should be taken when supplementing a medium with extra salts as this will change the osmolality. However, many cell lines have been shown to tolerate variation of.approximately 10% of this optimal value.

The salts that are primarily used arc those of Na^+, K^+, Mg^{2+}, Ca^{2+}, Cl^-, SO_4^{2-}, PO_4^{3-}, and HCO_3^-. These are present in most basal media formulations.

Trace Elements

Other inorganic elements that are present in serum in trace amounts may also be included in media formulations. These include Mn, Cu, Zn, Mo, Va, Se, Fe, Ca, Mg, Si, Ni. The less common elements that are not seen as essential have also been supplemented into media, including Al, Ag, Ba, Br, Cd, Co, Cr, F, Ge, J, Rb, and Zr. Many of these elements play an important role in enzyme activity and are essential to the survival of most cells.

The most commonly added element to serum-free formulations is selenium. The major functions of selenium can be attributed to its antioxidative properties and its role in cell growth. Selenium functions with mammalian cells in the form of selenoproteins, which have a number of physiological roles, of which 11 have been identified. Most of these proteins have been proven or implicated in antioxidant activities, such as glutathione peroxidase and thioredoxin reductase.

In some cases the supplementation of certain trace elements can decrease the requirement of certain growth factors in the medium. A ferrous salt, for example, can be used to replace transferrin as an iron carrier in many cell lines, such as B-lym-phocytes, MDCK, and human diploid fibroblasts.

Calcium has been shown to act as an important element involved in cell proliferation. It is important to many processes, including signal transduction, cell division, and cell adhesion. Calmodulin, a calcium-activated protein, regulates many serine/threonine kinases involved in cell division, in a calcium concentration dependent manner. Elevated calcium levels has been shown to be beneficial to some cell lines. These include keratinocytes, and human diploid fibroblasts. The reduction of calcium in culture media can be effective in reducing the degree of cell-cell adhesion. Cell clumping is a major problem for some production systems as this may reduce productivity and cell viability. Reducing the calcium in the medium can avoid this problem and is routinely done in many agitated-culture systems.

Vitamins

The vitamin and hormone components are present at relatively low concentrations and are utilized as cofactors, the requirements for which show considerable variations between cell lines. Thus, the content of these cofactors varies considerably among different media formulations. The requirement for extra vitamins is increased whenever the serum concentration is reduced. Cells that are limited in vitamins will vary in cell growth and survival but not in maximum cell density.

The vitamins included in each basal media formulation will vary depending on the cell line for which it was designed. The more basic media formulations, such as BME, designed for HeLa cells and mouse fibroblasts, contain biotin, folic acid, nico-tinamide, pantothenate, pyridoxal, riboflavin, and thiamine. These are typically supplemented with serum, which provides the other cofactor and vitamin requirements. The more complex formulations, such as F12, DMEM, or M199, are designed for serum-free formulations and contain a greater source of vitamins. M199, for example, contains all of the vitamins in BME but also includes vitamins A, B6, C, D3, E, K, and *p*-aminobenzoic acid (PABA). Each of these vitamin formulations was designed empirically for a specific cell line. Therefore, the requirements of another cell line may be different and will have to be determined independently.

Buffers

Bicarbonate is often included to act as a buffer system in conjunction with the carbon dioxide environment (5-10%) in which the cells should be cultured. This allows the cultures to be maintained at the normally optimum pH range of 6.9-7.4. The CO_2 is provided by a controlled atmosphere inside an incubator or alternatively in a sealed flask. The disadvantage of the bicarbonate-CO_2 buffer system is that the cultures may become alkaline very quickly when removed from the CO_2 incubator. To prevent this, the organic buffer Hepes (pH 7.0) may be added at a concentration up to ~25 mM. This allows the CO_2 level to be reduced to 2%.

Serum

Basal media used for the culture of mammalian cells are usually mixtures of a number of chemically defined components, such as vitamins, amino acids, salts, and sometimes protein supplements. Serum, which is a supernatant of clotted blood, is commonly used to provide animal cells with other components necessary for survival. These include attachment factors, micronutrients (trace elements, water-insoluble nutrients), growth factors (hormones, proteases), and protective elements (antitoxins, antioxidants, antiproteases). In most cases the growth requirements of the cell are not met until serum is added to the culture medium at a concentration of 5–10%.

Serum can be obtained from various animal sources—bovine or equine being the most common. One of the most effective supplements for cell growth is fetal calf serum because of its high content of embryonic growth factors.

Serum-free Media

The use of serum in culture media is associated with several problems including:

a. Batch to batch variation, which causes inconsistency in growth-promoting properties.
b. The high protein content, which can hinder product purification.
c. The risk of contaminants—viruses, mycoplasma, prions.
d. The risk of transmission of these contaminants to an end product used by humans, e.g., bovine spongiform encephalopathy (BSE) or mad cow disease.
e. Interfering with the effect of hormones or growth factors when studying their interaction with cells.
f. Limited availability of fetal calf serum, with periods of world shortages.
g. High cost of fetal calf serum, which can account for up to 85% of the overll cost of the medium when calculated for large-scale cultures.

Furthermore, serum-containing culture systems are becoming undesirable for large-scale processes in industry. The recent threat to human health caused by the undefined agents of BSE is likely to limit the continued use of bovine serum in culture processes used for the synthesis of health-care products such as viral vaccines.

For reduced serum, or serum-free formulations, various substitutes can be used. These include defined hormone cocktails such as the well-used HITES or ITES media, which contain hydrocortisone, insulin, transferrin, ethanolamine, and sele-nite. Alternatively, serum-free formulation may include growth factor extracts from endocrine glands such as epidermal or fibroblast growth factors. Every cell line is specific in its growth requirement and each serum-free formulation can only be applied to a limited number of cell types. Serum-free formulations have worked well for fast-growing tumorigenic or transformed cell lines, whereas their development for certain fastidious cell lines can be more difficult.

Since serum contains many undefined nutrients it has been difficult to reproduce the growth results associated with serum-supplemented media. Typical serum-free formulations have utilized a wide variety of components to mimic the action of serum. These have included purified proteins (animal or recombinant origin, see later), peptones, amino acids, inorganic salts, and animal cell or plant hydro-lysates. Although undefined in nature, they can provide significant enhancements to cell growth and proliferation. However, the disadvantages are similar to that of serum supplementation, such as contamination and inconsistent cell growth from one batch to another.

Serum-free media formulations have been developed for commonly used cell lines [such as hybridoma and Chinese hamster ovary (CHO)] as well as for some anchorage-dependent cell lines. The difficulty in the creation of formulations for some cell lines has been their fastidiousness and it is only recently that an understanding of growth factors and hormones has allowed the development of effective serum-free formulations. These lines have included those involved in cancer research, immunology, and even cell transplantation. In addition, there has been success with human lymphocytes, smooth muscle cells, and umbilical cord vein endothelial cells. A few examples of serum-free formulations are shown in Table 3.3.

Serum-free media for certain cancer cell lines, such as melanocytes, have been developed. In most cases a keratinocyte serum-free formulation such as MCDB 153 has been used. However, these cells often require specific growth factors, such as melanocyte-stimulating hormone (alpha-MSH) or dibutyryl cAMP, to induce proliferation. Otherwise the presence of normal keratinocytes or a keratino-cyte-conditioned medium is required to promote melanocyte cell growth. Other cancer cell lines that have serum-free formulations in the literature include prostatic carcinomas, lung carcinoma, bladder carcinoma (protein-free, serum-free medium).

The range of available serum-free media formulations and the demand for customized formulations are likely to increase as new cell lines are being established. Presently, there are still limited types of serum-free formulations available from commercial suppliers, and the selection still tends to be quite limited. Many claim to be "all-purpose" serum-free media to support a range of cell lines, but due to the variability of the growth factor requirement between cell lines, they remain quite limited in supporting cell growth. Further research in this area will result in the development of new and better formulations.

Chemically Defined Serum-Free Media

There is a clear distinction between serum-free media and chemically defined serum-free media. Chemically defined media require that all of the components must be identified and have their exact concentrations known. The supplements may be classified into several groups such as peptide growth factors, hormones, carrier proteins, hydrolysates, and attachment factors. The advantages of having chemically defined media are the same as those listed previously. It also allows researchers who are studying in the field of cell physiology (especially extracellular) and/or molecule-cell interactions to eliminate any variables that may arise due to the effects of unknown components in the medium.

The advantages of using media supplemented with defined growth factors and hormones include the following:

a. In many cases cells show enhanced growth characteristics in these defined media compared to serum-containing media. Some cells, notably differentiated types, cannot be maintained at all in serum-supplemented media and their maintenance in vitro is dependent on selected hormones and growth factors.

TABLE 3.3 EXAMPLES OF SERUM-FREE FORMULATIONS DESIGNED FOR VARIOUS CELL LINES

	Molarity (M)				
Components	Hematopoietic progenitor cells	MDCK	Lung carcinoma	CHO	Human umbilical endothelial
Inorganic salts					
Calcium chloride ($CaCl_2$)	1.49×10^{-3}	1.05×10^{-3}	1.05×10^{-3}	8.93×10^{-4}	
Calcium nitrate ($Ca(NO_3)_2 \cdot 4H_2O$)		1.20×10^{-10}	4.24×10^{-4}	1.20×10^{-10}	
Cupric sulfate ($CuSO_4 \cdot 5H_2O$)		7.80×10^{-12}		7.80×10^{-12}	5.01×10^{-9}
Ferric nitrate ($Fe(NO_3)_3$ $9H_2O$)		1.20×10^{-7}		1.20×10^{-7}	
Ferrous sulfate ($FeSO_4 7H_2O$)	5.00×10^{-6}	1.50×10^{-6}		1.50×10^{-6}	1.50×10^{-6}
Potassium chloride (KCl)	4.44×10^{-3}	4.16×10^{-3}	5.30×10^{-3}	4.16×10^{-3}	3.71×10^{-3}
Potassium nitrate (KNO_3)	7.52×10^{-4}				3.76×10^{-7}
Magnesium chloride ($MgCl_2$ anhydrous)					3.01×10^{-4}
Magnesium sulfate ($MgSO_4$)	8.14×10^{-4}	4.07×10^{-4}	4.07×10^{-4}	4.07×10^{-4}	
Magnesium sulfate ($MgSO_4$ anhydrous)					2.41×10^{-3}
Mangenous sulfate ($MnSO_4 \cdot H_2O$)	1.00×10^{-7}				1.46×10^{-6}
Ammonium molybdate ($(NH_4)6Mn_7O_{24} \cdot 4H_2O$)	1.00×10^{-7}				
Nickelous chloride ($NiCl_2 \cdot 6H_2O$)	5.00×10^{-8}				
Sodium chloride (NaCl)	7.76×10^{-2}	1.21×10^{-1}	1.03×10^{-1}	1.21×10^{-1}	1.04×10^{-1}
Sodium bicarbonate ($NaHCO_3$)	3.60×10^{-2}	0.02	2.38×10^{-2}	2.90×10^{-2}	2.50×10^{-2}
Sodium phosphate ($Na_2HPO_4 \cdot H_2O$)		5.00×10^{-4}		5.00×10^{-4}	5.00×10^{-4}
Sodium phosphate ($NaH_2PO_4 \cdot H_2O$)	9.06×10^{-4}	4.53×10^{-4}	5.63×10^{-3}	4.53×10^{-4}	4.53×10^{-4}
Stannous chloride ($SnCl_2$)	5.00×10^{-8}				
Sodium selenite ($Na_2SeO_3 \cdot 5H_2O$)	6.46×10^{-8}		3.00×10^{-8}	1.00×10^{-7}	4.79×10^{-8}
Ammonium metavanadate ($NaVO_3$)	5.00×10^{-7}				
Zinc sulfate ($ZnSO_4 \cdot 7H_2O$)		1.50×10^{-6}		1.50×10^{-6}	1.35×10^{-5}
Energy metabolism					

D-Glucose	2.50×10^{-2}	1.75×10^{-2}	1.11×10^{-2}	1.75×10^{-2}	1.75×10^{-2}
Sodium pyruvate	1.00×10^{-3}	5.00×10^{-4}		5.00×10^{-4}	5.00×10^{-4}
Nucleic acid derivatives					
2′-Deoxyadenosine	0.1mg/mL				
2′-Deoxycytidine	0.1mg/mL				
2′-Deoxyguanosine	0.1mg/mL				
Hypoxanthine (Na)		1.50×10^{-5}		7.35×10^{-5}	
Thymidine		1.50×10^{-6}		4.13×10^{-5}	1.51×10^{-6}
Adenosine	0.1mg/mL				
Cytidine	0.1mg/mL				
Uridine	0.1 mg/mL				
Guanosine	0.1mg/mL				
2′-Deoxythymidine	0.1mg/mL				
Amino acids					
β-Alanine	2.81×10^{-4}				
L-Alanine	3.98×10^{-4}	5.00×10^{-5}		1.00×10^{-6}	2.52×10^{-4}
L-Arginine-HCl	1.89×10^{-4}	7.00×10^{-4}	1.10×10^{-3}	1.40×10^{-3}	1.04×10^{-3}
L-Asparagine	2.30×10^{-4}	5.00×10^{-5}	3.79×10^{-4}	1.06×10^{-4}	2.71×10^{-4}
L-Aspartic acid		5.00×10^{-5}	1.50×10^{-4}	1.03×10^{-4}	2.53×10^{-4}
L-Cystine-2HCl	3.81×10^{-4}	1.00×10^{-4}		2.00×10^{-4}	1.46×10^{-4}
L-Cysteine-HCl·H_2O		1.00×10^{-4}	2.06×10^{-4}		1.00×10^{-4}
L-Cystine	5.10×10^{-4}				
L-Glutamic acid	4.00×10^{-3}	5.00×10^{-5}	1.36×10^{-4}	9.14×10^{-5}	5.43×10^{-4}
L-Glutamine	2.00×10^{-3}	2.50×10^{-3}	2.05×10^{-3}	5.00×10^{-3}	9.00×10^{-3}
Glycine	4.00×10^{-2}	2.50×10^{-4}	1.33×10^{-4}	5.00×10^{-4}	2.50×10^{-4}
L-Histidinehydrochloride·H_2O		1.50×10^{-4}		1.50×10^{-4}	2.19×10^{-4}
L-Histidine			9.67×10^{-5}	4.00×10^{-4}	
L-Hydroxyproline	8.02×10^{-4}		1.53×10^{-4}		
L-Isoleucine	8.02×10^{-4}	4.16×10^{-4}	3.82×10^{-4}	8.32×10^{-4}	4.15×10^{-4}
L-Leucine	7.98×10^{-4}	4.51×10^{-4}	3.82×10^{-4}	1.35×10^{-3}	4.50×10^{-4}
L-Lysine-HCl	2.01×10^{-4}	4.99×10^{-4}	2.19×10^{-4}	1.50×10^{-3}	6.53×10^{-4}

TABLE 3.3 EXAMPLES OF SERUM-FREE FORMULATIONS DESIGNED FOR VARIOUS CELL LINES (*CONTINUED*)

Components	Molarity (M)				
	Hematopoietic progenitor cells	MDCK	Lung carcinoma	CHO	Human umbilical endothelial
L-Methionine	4.00×10^{-4}	1.16×10^{-4}	1.01×10^{-4}	2.30×10^{-4}	1.34×10^{-4}
L-Phenylalanine	3.48×10^{-4}	2.15×10^{-4}	9.09×10^{-5}	4.27×10^{-4}	4.69×10^{-4}
L-Proline	4.00×10^{-4}	1.50×10^{-4}	1.74×10^{-4}	4.45×10^{-4}	3.24×10^{-4}
L-Serine	7.80×10^{-5}	2.50×10^{-4}	2.86×10^{-4}	2.50×10^{-4}	7.26×10^{-4}
L-Threonine	7.80×10^{-5}	4.49×10^{-4}	1.68×10^{-4}	8.94×10^{-4}	4.49×10^{-4}
L-Tryptophan	4.62×10^{-4}	4.42×10^{-5}	2.45×10^{-5}	1.32×10^{-4}	4.42×10^{-5}
L-Tyrosine-2Na·$2H_2O$	8.03×10^{-4}	2.14×10^{-4}	1.10×10^{-4}	5.23×10^{-4}	2.49×10^{-4}
L-Valine		4.52×10^{-4}	1.71×10^{-4}	9.05×10^{-4}	5.54×10^{-4}
Vitamins					
Ascorbic acid	5.32×10^{-8}				1.00×10^{-3}
Biotin	8.30×10^{-6}	1.43×10^{-8}	8.00×10^{-6}	1.43×10^{-8}	4.15×10^{-8}
D-Calcium pantothenate	2.85×10^{-5}	4.60×10^{-6}	5.00×10^{-7}	4.60×10^{-6}	4.70×10^{-6}
Choline chloride	9.06×10^{-6}	6.41×10^{-5}	2.14×10^{-5}	6.41×10^{-5}	6.43×10^{-5}
Folicacid	4.00×10^{-5}	6.00×10^{-6}	2.20×10^{-6}	6.00×10^{-6}	6.00×10^{-6}
i-Inositol		7.00×10^{-5}	1.94×10^{-4}	7.00×10^{-5}	6.99×10^{-5}
Nicotinic acid (Niacin)	3.28×10^{-5}				
Nicotinamide		1.65×10^{-5}	8.10×10^{-6}	1.65×10^{-5}	1.65×10^{-5}
P-aminobenzoic acid	1.96×10^{-5}		7.20×10^{-6}		
Pyridoxal-HCl	1.06×10^{-6}		4.80×10^{-6}		9.82×10^{-6}
Pyridoxine hydrochloride		9.70×10^{-6}		9.70×10^{-6}	1.51×10^{-7}
Riboflavin	1.18×10^{-5}	6.00×10^{-7}	5.00×10^{-7}	6.00×10^{-7}	5.82×10^{-7}
Thiamine-HCl	9.59×10^{-9}	6.40×10^{-6}	2.90×10^{-6}	6.40×10^{-6}	6.43×10^{-6}
Vitamin B_{12}		5.00×10^{-7}	3.69×10^{-9}	5.00×10^{-7}	5.06×10^{-7}
Lipids and derivatives					
Cholesterol	2.00×10^{-6}				
i-Inositol			1.9×10^{-4}		

Linoleic acid (mg/mL)		1.5×10^{-4}		1.5×10^{-4}	42
Lipoicacid		5.1×10^{-4}		5.1×10^{-4}	105
Ethanolamine (mg/L)				6	
Prostaglandin E_1 (ng/mL)		25			
Phophatidylcholine (mg/L)				5	
BSA/oleic acid (mg/L)					200/1.6
Other compounds					
EGF (ng/mL)	10				.10
Putrescine-2HCl		5.03×10^{-7}		5.03×10^{-7}	5.00×10^{-7}
Glutathione (reduced)			3.20×10^{-6}		
Hydrocortisone		5.00×10^{-8}	10	2.76×10^{-5}	2.76×10^{-6}
Insulin (mg/mL)		5	5	2	5
BSA	1% w/v				
2-Mercaptoethanol	5.00×10^{-5}				1.25×10^{-5}
Transferrin (mg/mL)	0.7	5	100	10	5
Stem cell factor (ng/mL)	10				
Interleukin-1β (ng/mL)	3				
Interleukin-3 (ng/mL)	100				
Interleukin-6 (ng/mL)	100				
Erythropoietin (U/mL)	1				
Triiodothronine		5.00×10^{-12}			
Estradiol (17-β)			1.00×10^{-8}		
βFGF (ng/mL)					10
Liver growth factor (gly-his-lys) (mg/mL)					50
VEGF (ng/mL)					5
N-Acetyl-l-cysteine					2.50×10^{-4}
Buffers/indicators					
Hepes	25	15			1.25×10^{-2}
Phenol red	0.0346	2.04×10^{-5}	1.25×10^{-5}	2.04×10^{-5}	1.21×10^{-5}

b. It is possible to select specific cell types from a mixed population of cells as may be obtained from a primary source. Such cell-specific selection requires careful manipulation of the medium composition.
c. Studies of the interaction of hormones or drugs on cultured cells are made possible by such media. The uncertain composition of serum and the binding effects of the associated proteins would limit such studies in serum-supplemented media.

Some examples of serum-free formulations that are not chemically defined are those that contain various serum substitutes, such as serum fractions or animal/plant hydrolysates. Although such serum-free media are more defined than serum-containing media, there remains the risk of introducing harmful contaminants such as viruses. Therefore, the only way to ensure a safe serum-free medium is to completely replace any animal-derived substances by non-animal-derived products, such as those of plant origin.

Growth Factors

This section includes information on some of the specific proteins used as supplements, which are often added to cell culture media in order to enhance cell growth. These components are supplements to basal media that will promote growth in the absence of serum or in reduced serum formulations.

Many polypeptides have been characterized as mitogenic in vitro for specific mammalian cell lines. This family includes fibroblast growth factor (FGF: aFGF and bFGF), insulin-like growth factor (IGF), epithelial growth factor (EGF), nerve growth factor (NGF), platelet-derived growth factor (PDGF), and transforming growth factor (TGF: a and b). These growth factors are active in the 1–10 ng/mL range.

Fibroblast Growth Factor

FGFs are pleiotropic growth factors that control cell proliferation, migration, and differentiation in a wide variety of cell types. They exist as two types: acidic FGF (p*I* 5.6) and basic FGF (p*I* > 9.0). Due to their affinity to heparin they are sometimes called heparin-binding growth factors, although only the activity of acidic FGF is altered by heparin.

Acidic FGF

This form of FGF carries 55% homology with basic FGF. Also known as endothe-lial growth factor, this is the less potent of the two mitogens. In the absence of heparin sulfate on the surface of target cells, or free heparin, FGF cannot exert its biological activity due to its instability and is easily damaged by proteolysis. For this reason it is often supplemented with heparin. In some cases synthetic dextrans have been used as an alternative to heparin.

Basic FGF

This is produced by most cell types investigated in vitro, and the FGF receptor is ubiquitous. Thus, bFGF can act as both an autocrine (stimulation of the same cell) and a paracrine growth factor (one cell stimulating an adjacent cell). Basic FGF stimulates cell proliferation of a wide variety of cell types through activation of protein kinase C.

It is a potent mitogen for many types of cultures derived from mesodermal cells, neuroectoderm, and several transformed cell lines. It also promotes the differentiation of adipocytes and ovarian ganulosa cells, and induces the synthesis of extracellular matrix (ECM) proteins such as collagen type IV.

Epithelial Growth Factor

EGF is a potent mitogen for a variety of cell lines including primary, mesenchymal, epidermal, and glial cells. Cells that are transformed tend to lose their requirement for EGF. The activity of EGF is linked to its activation of a transmembrane tyrosine kinase, which plays a role in cellular signaling and development. Many of its activities include the induction of gene expression, changes in ion flux, and mitogenesis. Its mitogenic activity is synergistic with that of other peptide growth factors such as IGF-1 and TGF.

Nerve Growth Factor

NGF is a well-characterized molecule required for the survival and differentiation of a variety of cell types both in the peripheral and in the central nervous system. It is not a potent mitogen, but it does induce the differentiation and enhances the survival of sympathetic neurons and PC 12 phaeochromocytoma cells in culture. NGF initiates the majority of its biological effects by activation of the tyrosine kinase receptor

TrkA. These signals are then propagated to other messengers involved during neuronal differentiation.

Transforming Growth Factor

Transforming Growth Factor Alpha

This factor bears 30% homology to EGF, and binds to EGF receptors with equal potency. Its actions are similar to those of EGF and it is a potent epithelial oncopro-tein. Usually, a culture will not require EGF supplementation if *TGFα* is present. TGFα is synthesized as a large membrane-bound precursor, which is subsequently cleaved by proteases into the mature peptide. TGFα is found naturally in the embryonic kidney, adult brain, pituitary gland, skin, and placenta.

Transforming Growth Factor Beta

Three isoforms of TGFβ (β1, β2 and β3) exist in mammals. They play a critical role in growth regulation and development. All three growth factors are secreted by most cell types, generally in a latent form, requiring activation before they can exert biological activity. Activation of TGFβ requires an acidic environment or the presence of certain activators, such as plasmin or thrombospondin.

The highest concentrations of TGFβ1 are found in platelets and bone, with smaller amounts expressed by many cell types. It is closely conserved between species with ≈98% equivalency. Mature TGFβ1 in human, porcine, chicken, simian, and bovine species are identical, and differ from murine TGFβ1 in a single amino acid position.

The effects of TGFβ on proliferation and differentiation can be stimulatory or inhibitory. This is dependent on many factors, including cell type, growth conditions, the state of cell differentiation, and the presence of other growth factors. TGFβ promotes the growth of several mesenchymal cell lines in soft agar (e.g., fibroblasts) in combination with either TGFα or EGF. In contrast, TGFβ alone *inhibits* the growth of many cell lines in monolayer culture, which may result from its ability to stimulate the secretion of ECM proteins (collagens, fibronectin, and glycosamino-glycans) and protease inhibitors. TGFβ can inhibit monolayer growth in epithelial cells, endothelial cells (antagonized by bFGF), stem cells, and lymphocytes.

Insulin and IGFs

Insulin

Several of the mitogenic actions previously ascribed to insulin may be due to IGF contamination of the original preparations. Insulin is quite unstable at 37°C (particularly in media with high levels of cysteines) with >90% of its activity being destroyed over 1 hr. It is often added to media at relatively high concentrations and because insulin has a weak affinity for IGF-1 receptors, it may activate the same mitogenic responses as IGF-1 at these high doses. However, insulin is also added to most serum-free media for its ability to promote energy and anabolic metabolism (e.g., glucose uptake and oxidation, synthesis of glycogen, amino acid transport). The ubiquitous insulin receptor is a dimer that has tyrosine kinase activity. Insulin requires adequate zinc to exert its biological function and should be present in the culture medium.

Insulin-Like Growth Factors

These peptides are homologous to insulin and evolved by gene duplication. They were first discovered as the serum components that mediate the effects of growth hormone on cartilage—hence, their former name of somatomedins. IGF activity is regulated by a family of six IGF-binding proteins that circulate in the blood. They extend the life and modulate the activity of IGF on various tissues. Many of the metabolic actions of IGF overlap with that of insulin; however, they are more potent mitogens and synergize with FGF, EGF, and PDGF.

Insulin-Like Growth Factor 1. IGF-1 is secreted in low amounts by most tissues, and acts on a wide range of mesenchymal cells. Although IGF-1 binds to the insulin receptor with low affinity, its main actions are mediated by a specific IGF-1 receptor, which has intrinsic tyrosine kinase activity. This receptor also binds IGF-2 and, to a lesser extent, insulin.

Insulin-Like Growth Factor 2. Formerly known as multiplication-stimulating activity, this factor is produced by a more restricted range of tissues (e.g., fetal liver, muscle, skin, and adult brain). It exhibits insulin-like activity, but it is not as potent as IGF-1. Its receptor has lower affinity to insulin or IGF-1. This receptor is found on adipocytes and (in contrast to insulin and IGF-1 receptors) does not display tyrosine kinase activity.

Platelet-Derived Growth Factor

PDGF is a major mitogen for fibroblasts, smooth muscle cells, and other cells. It was originally identified as a component of whole-blood serum (40-60 ng/mL) that was subsequently isolated from human platelets.

PDGF consists of two structurally similar polypeptide chains (A and B), which combine to form homo- and heterodimers. The three-dimensional structure of PDGF-BB has been shown to have similarity to NGF, TGFβ, and vascular endothe-lial growth factor (VEGF)s, which have homologous amino acid sequences.

The PDGF isoforms promote a response by binding to two structurally similar binding sites of a protein tyrosine kinase receptor, designated α and β. Binding of PDGF can result in mitosis, chemotaxis, actin reorganization, or prevention of apoptosis in certain cell lines.

In serum both forms of PDGF bind to α_2-macroglobulin. PDGF synergizes with EGF and IGF-1 in stimulating normal Balb/C 3T3 fibroblasts to proliferate. Its action is restricted to mesenchymal and neuroectoderm cells, which have specific PDGF receptors exhibiting tyrosine kinase activity.

Interleukin-6

Interleukin-6 (IL-6) is a cytokine that regulates many immune and hematopoietic systems. Its function is pleotropic and it exhibits many overlapping biological functions with other cytokines such as IL-11, leukemia inhibitory factor, oncastatin, ciliary neurotrophic factor, and cardiotrophin-1, as they share a common signal transducing receptor, gp130. It does not stimulate tyrosine kinase activity directly, but rather it stimulates other cytoplasmic tyrosine kinases and subsequent modification of transcription factors.

It is produced by many cell types including macrophages, fibroblasts, T-cells, and endothelial cells. Its effect upon some cell lines can be proliferative and mito-genic. Treatment of the PC 12 cell line with IL-6 causes differentiation and it can also protect them from cell death under serum deprivation. It is also mitogenic to other cell types such as stem cells and hybridomas. Similarly, IL-6 can stimulate B-cells to proliferate in vitro. Many of these effects can be synergistically enhanced with simultaneous treatment with other growth factors such as EGF, FGF, or NGF.

Not all responses to this cytokine are proliferative. Certain hybridomas increase their antibody production when treated with IL-6. In addition, IL-6 synergizes with other ILs to enhance antibody production.

In cell culture, IL-6 can replace the use of feeder cells (such as peritoneal exu-date cells, spleen cells, fibroblasts, or thymocytes) during cloning by limiting dilution and in the post-fusion stages of hybridoma culture.

Reconstitution and Storage of Growth Factors for Serum-Free Media

Many of the components of serum-free formulations are not particularly stable in a protein-free environment. These include recombinant proteins such as IGF, EGF, PDGF, FGF, TGF, and others. These compounds should be reconstituted in water with a protein carrier, such as bovine serum albumin (BSA), at a high concentration (~mg/mL). The solution should be kept frozen at <–20°C depending on the instructions supplied with the product information sheet. The solution can then be thawed and introduced into the medium immediately before being used for cell culture, or even directly into the culture flask.

Many of the compounds may not require a protein carrier upon reconstitution (if animal components are not desirable in the medium). They can be dissolved in water, sometimes with a little acid at low concentration (10 mM). However, the absence of a carrier will reduce the stability of the recombinant protein and the compound may only be stable at <–20°C for a few months.

Lipids

Lipids serve a variety of functions within the cell. They act as a structural component for cell membranes, in sensors for external signals, and as a source of metabolic energy. Serum itself contains lipids, but basal media formulations do not normally contain any lipid components. Certain lipids can be included as supplements to serum-free formulations and have been shown to enhance cell proliferation. The types of lipids commonly found in serum include fatty acids, phospholi-pids, lecithin (phophatidylcholine), and cholesterol. Phospholipids are not only major components in cell membranes but also play an important role in signal trans-duction pathways regulating growth. When introduced extracellularly, phospholipids have shown growth stimulatory effects on many cell lines. Phosphatidic acid and lysophosphatidic are growth promoting for some anchorage-dependent cells, such as MDCK, mouse epithelial, and other kidney cell lines. Others, such as phosphatidylcholine, phosphatidylethanolamine, and phosphatidylinositol, stimulate the growth of human diploid fibroblasts in serum-free formulations. Ethanolamine is growth stimulatory for some hybridoma cell lines, and is often supplemented to serum-free formulations. Cholesterol has also been shown to be stimulatory for most mammalian cell lines and is a requirement for growth of certain cells, such as myelomas. Other specific fatty acids, such as oleic and lino-leic, will enhance both the growth and the productivity of hybridomas grown in a serum-free medium over an extended number of passages.

Carrier Proteins

These are important to promote the transport of essential nutrients or trace elements. A brief description of some of the carrier proteins follows.

Bovine Serum Albumin

BSA is commonly found in circulating plasma and plays an important role in cell culture. It is primarily used as a carrier of lipids, such as linoleic and oleic acids, since they are insoluble in aqueous solution. In addition to fatty acids it also has a high affinity for other compounds, such as various metals (Fe, Cu, and Ni). It also serves as a protective agent of toxic metabolites such as oxygen-free radicals and bilirubin. In addition, BSA is a transport protein for nitric oxide, which is important for neurotransmission.

In some cases BSA is undesirable for protein-free and serum-free formulations. In these cases alternative non-animal-derived compounds can be used as lipid carriers, such as cyclodextrin.

Transferrin

This is an important iron carrier protein and is essentially the source for all circulating plasma iron. The function of transferrin is to solubilize the iron, prevents free-iron toxicity, and facilitate the transport of iron into cells. Iron uptake occurs as an iron-transferrin complex and is internalized by receptor-mediated endocytosis. For the purposes of culture media, human transferrin is more commonly used than bovine transferrin, due to its ability to deliver iron to a wider range of cell types.

ECM PROTEINS

A number of proteins are important to ensure the attachment of anchorage-dependent cells to their substratum. The attachment proteins (also referred to as ECM proteins) are present in serum and can also be secreted by the cells. Some adherent cells do not bind directly onto plastic and require ECM proteins for efficient plating and growth. Some of the commonly used ECM proteins in cell culture will be discussed.

Fibronectin

Fibronectin is a high-molecular weight multimeric glycoprotein found on the surface membrane of cells. There are two forms of fibronectin that are generally available. Plasma fibronectin has been isolated from blood plasma and plays a role in blood clotting, wound healing, and phagocytosis.

In cell cultures, fibronectin is important in cell-substratum adhesion of anchorage-dependent cells. As cells synthesize fibronectin, the need for media supplementation is dependent on the cell type. Transformed cells lose the ability to produce fibronectin and this may partly explain their loss of anchorage dependence. Many cell types show improved adherence and growth on fibronectin-coated surfaces. Furthermore, in the absence of serum, fibronectin has been shown to enhance the survival of primary fibroblasts and endothelial cells in culture by suppressing p53-regulated apoptosis via transduction signals through focal adhesion kinase.

In addition to adhesion, fibronectin plays an important role in stimulating many mitotic processes (such as the mitogen-activated protein kinase pathway) and has been shown to stimulate cell cycle progression in nontransformed cell lines, such as NIH-3T3 and BHK.

Laminin

Laminin is another ECM protein and is sometimes used as an alternative to fibronectin, especially for plating epithelial cells. It is similar in structure to fibronectin, but more predominant in differentiated tissues, and is one of the major components of the basal lamina. It also plays an important role in the regulation of neuronal migration, being a potent promoter of cell outgrowth in cultures, especially neural cell lines. In conjunction with a polylysine peptide, laminin has been shown to provide guidance for neuron attachment and axon outgrowth under serum-free conditions.

Pronectin

Pronectin is a recombinant peptide designed to replace fibronectin in serum-free media formulations. Similar in structure to fibronectin, it contains the same repeats of the RGD (Arginine-Glycine-Aspartate) cell attachment sequence. It has been used to promote attachment of various anchorage-dependent cells in both stationary and microcarrier cultures, such as the MDCK cell line.

Choice of Supplements for Serum-free Media Formulations

Deciding which supplements are important for each cell line can be the most difficult task during development of a new serum-free medium. Rather than using classical approaches to development (e.g., Ham and Sato), using formulations that have been developed for similar cell lines can serve as a guide, or at least a starting point, in creating a new formulation. However, the growth factor requirements for any given cell line can vary dramatically from one another, so this approach does not always work.

There are many growth factors to choose from, and the effect of each may be cell line dependent. A brief guide, based on cell type, follows:

a. *Epithelial cells:* Growth factors that are often included are EGF, corticosteroids, and retinol. ECM proteins can also be benefi-cial, including fibro-nectin, laminin, and collagen, to promote cell adhesion.
b. *Mesenchymal cells:* Growth factors that are often included are acidic/basic FGF (basic being more potent than acidic), EGF (or TGFα),IGF/insulin (but IGF being more potent), PDGF (BB most potent), and sometimes TGFβ (but it is growth inhibitory for most cells). Other supplements can include transferrin, corticosteroids, ECM proteins such as fibronectin, and trace elements.
c. *Transformed cells:* These cells often have a lower requirement for growth factors. Supplements used have included IGF/ insulin, carrier proteins BSA/fetuin, human transferrin, and trace elements.

Iron Salts as a Replacement for Transferrin

Transferrin is used in culture media as an iron transporter for animal cells. However, for protein-free media or for media that do not contain components of animal origin, transferrin can be replaced by simple ferric salts. This has been accomplished in many media formulations for various cell lines, such as hybridoma. Ferrous salts tend to oxidize quickly, however, so the medium should be used within a few weeks or they should be introduced just before the medium is to be used. Typical sources of iron in serum-free formulations have been ferrous sulfate or ferric citrate.

Undefined Serum Substitutes

In an attempt to further enhance the performance of serum-free formulations, various extracts of serum or plasma have been developed. These fractions may serve to replace the growth-promoting effects of whole serum. Although a serum-free medium containing these fractions is not chemically defined, it confers the advantage of lowering the protein content, which in turn can reduce the batch-to-batch variability. The only disadvantage is that the exact chemical composition is not known.

Cohn fractionation of blood is a method used to separate plasma fractions by sequential steps of ethanol precipitation. This method is used for the large-scale preparation of hemophilic factors from pooled human blood. Cohn fraction 4.1, which is a by-product of this process, has been found to possess growth-promoting properties that can substitute for whole serum in growth medium, and be used to promote the growth of MDCK cells.

Commercial sources of these substitutes are available and are designed as complete serum-free media for cell culture. These media formulations are usually tested on specific cell lines in comparison to serum-based media in growth-promoting ability. Some of the companies that supply these serum-substitutes are Invitrogen (Gibco), Hyclone, and Sigma.

Other types of substitutes have also been developed. Hydrolysates, which are enzymatic digests of proteins, are being widely used in cell culture. They are obtained from sources such as animal tissue, milk, and plants such as soy, wheat, or rice. The end product is a source of vitamins, lipids, minerals, and di- and tripeptides that supply a rich energy source of basic amino acids for the cell.

Hydrolysates are typically used as a supplement to a basal medium, although they can be used to replace it altogether. The types of cell lines that have been propagated with hydrolysates include CHO, hybridoma, BHK, Vero, lymphocytes, and others.

The advantages of using hydrolysates to replace serum include the following:

a. There is no excess protein, which can interfere with purification.
b. Any hormones that were introduced with serum are no longer present.
c. There is a significant reduction in the cost of the media.
d. Utilizing plant hydrolysates will decrease the risk of introducing animal-derived contaminants into the cell culture process.

Although serum-free formulations that contain these hydrolysates confer certain advantages, the problem of batch variation still remains. Furthermore, hydrolysates derived from animal tissue can still carry the risk of introducing contaminants, such as viruses or prions, into a cell culture process used in the production of biologicals. This can limit their use in the large-scale production of human therapeutics due to health regulatory concerns.

Production of Biologically Active Substances by Serum-Free Cultures

Animal cell technology has allowed the production of a great diversity of biological substances including monoclonal antibodies, vaccines, hormones, and other regulatory molecules. The development of serum-free media has improved these processes by conferring many advantages as discussed previously.

The vaccine industry has seen the emergence of new processes for production utilizing serum-free culture systems. In many cases the traditional method of production has led to many problems including adverse allergic reactions (e.g., from chick eggs used to propagate the virus) to the transfer of biological contaminants from animal components used in the process. Furthermore, many of the more traditional processes were inefficient and time-consuming. The use of animal cells for the production of vaccines alleviated some of these problems by allowing large-scale production with minimal supervision and monitoring. However, the use of serum in the culture medium was still required, which posed the threat of transferring contaminants to the end product. The recent advancements in serum-free formulations have reduced the dependence on serum in industrial-scale processes. This has allowed the minimization of potential contamination by adventitious agents.

It has been demonstrated that serum-free cultures can have comparable productivity to their serum-supplemented counterparts and have been widely accepted in industry. Some examples of vaccines produced today under serum-free conditions are rabies, polio, and influenza. Other important viruses, such as HIV, adenovirus, and reovirus (to name a few) have also been produced under serum-free conditions.

Serum-free culture systems have also been used for the production of other biologically active substances. Cytokines, which include the lymphokines and interfer-ons, are of particular interest to industry due to their involvement in the immune and inflammatory responses within the body. IL-2, for example, has been produced from BHK cells under serum-free conditions employing both stirred tank reactors and a hollow fiber system. The yields of the system ranged from 0.75 mg/L in the stirred tank system to 0.23 mg/L in the hollow fiber module, where the cell densities reached 3.0 × 107 cells/mL and 6.0 × 107 cells/mL, respectively. Another group discovered that IL-2 production between serum-containing and serum-free cultures resulted in identical titers.

The production of interferon has been achieved in a wide variety of cell lines under serum-free conditions. Gamma-interferon, for example, normally obtained from T- or B-lymphocytes, has been produced using CHO cells. In this case certain components of the serum-free medium were important for interferon production and affected the productivity of the culture system. They included BSA, sodium pyruvate, glutamate, methionine, proline, histidine, hydroxyproline, tyrosine, and phenylalanine. Similarly, the production of beta-interferon in Namalwa cells under serum-free conditions has been studied.

Other biologicals, such as tissue plasminogen activators (tPA), blood clotting factors, hormones, and polypeptide growth factors have been produced in vitro in serum-free cultures. Some examples of these processes include the production of tPA. In this serum-free system, cultures of fibroblasts on Cytodex-3 microcarriers gave yields comparable to that of 5% fetal bovine serum-containing media. Also, a serum-free culture of CHO cells in the production of recombinant human growth hormone allowed the elimination of a final purification step. The elimination of serum has allowed continuous nutrient optimization to improve the productivity.

The importance and wide-spread use of monoclonal antibodies in the pharmaceutical industry has led to the development of better serum-free formulations for large-scale production. They are becoming a standard for industry due to the high reproducibility of growth conditions, the ability to optimize nutrient requirements, and the reduction of possible contaminants. Furthermore, hybridoma cells grow well in suspension culture and are relatively easy to adapt to serum-free formulations compared to anchorage-dependent cell lines, which tend to be more fastidious. For these reasons they are commonly adapted to serum-free media for large-scale production systems.

In most cases the conversion to serum-free media has led to improvements for the biopharmaceutical industry, including increased productivity, efficiency, and cost-effectiveness.

Protein-free Media

Protein-free media are typically serum-free formulations in which the protein growth factors or other protein substitutes have been removed. These can be in the form of BSA, fetuin, or others such as transferrin. This can create problems in downstream product purification by necessitating an increase in purification steps. As a consequence, lower product yields or a decrease in purity can result. Therefore, protein-free media can greatly increase the productivity and efficiency of processes for producing biologicals.

The elimination of serum alone reduces the amount of protein in the medium. However, the removal of proteins such as transferrin, insulin, and fetuin from a serum-free formulation can drastically reduce the performance of the medium. To counter these effects other supplements can be used, such as animal tissue or plant hydrolysates. These compounds have been shown, in some cases, to increase the growth rate of cell lines in protein-free formulations compared to that of serum-containing media. Since these hydrolysates do not contain any protein they do not interfere with the purification procedures in downstream processes.

In many cases the yields obtained from protein-free cultures are higher than those of their serum-containing counterparts. For monoclonal antibody production an increase in productivity has been observed for many cell lines, including hybri-doma, CHO, and BHK. Similar results have also been observed for the production of viruses, such as HIV.

Strategies For The Development Of Serum-free Media

There are many analytical approaches for designing serum-free formulations. The objective is to identify growth-promoting components that may be present in serum and include them in a completely synthetic medium.

Ham's Approach

This method is a systematic approach to the development of serum-free media by the gradual reduction in the concentration of undefined supplements in the culture. It is a time-consuming but successful method to develop a single formulation for a given cell line.

For nutrients in the basal medium there is a wide concentration range over which the nutrient is not limiting. Starting with a serum-containing medium, a range can be established by adding the nutrient at concentrations of 0.1 ×, 1×, 10×, 100×, etc., of the original concentration. By comparing the growth response at each concentration, the optimum concentration range can then be determined through the construction of a growth response curve. The optimum concentration for the nutrient is set near the center of the broad optimum plateau to reduce the likelihood that it will become growth limiting through the adjustments of other medium components. In the response curve shown in Fig. 1 the center of the optimum concentration is 10–5M.

This procedure of optimization of the basal medium may reduce the dependence of serum supplementation to a low level (2%). The cells can adapt to reduced serum by the stimulation of anabolic processes. It may be necessary to allow the cells to grow for several generations at each step of serum reduction to allow the required adaptation to become established.

The ability of cells to grow in a serum content of <2% will vary and may depend on the availability of growth factors and hormones to replace those normally supplied by the serum. A dialysis step is useful to distinguish between the requirements for low- and high-molecular weight factors. The addition of selected components to cells growing at 50% maximal rates provides for easy methods of detecting growth enhancement.

Steps in Ham's approach to the design of serum-free media:

a. Use any means necessary to obtain growth of the cells of interest. This may include serum supplementation, the use of feeder layers or of conditioned media.
b. Select the combination of readily available media to obtain the maximum cell growth.

c. Replace all undefined supplements with dialyzed supplements. Add the undefined low-molecular weight supplements that were removed by dialysis.
d. Identify all the low-molecular weight supplements required for growth by a combination of analysis and trial additions.
e. Reduce the addition of dialyzed supplements to a level that supports less than maximal growth (for example, 50% of maximum). Thus, the undefined supplements are now rate limiting to cell growth.
f. Sequentially adjust the concentrations of all defined nutrient components of the basal media to experimentally determined optima.
g. Sequentially test defined low-molecular weight substances for growth enhancement.
h. At each stage of growth improvement, reduce the amount of undefined supplement so that it becomes rate limiting.
i. Repeat the procedures in steps (g) and (h) until no further reduction in the amount of dialyzed supplement is possible without affecting cell growth.

Sato's Approach

Sato and colleagues have developed an approach based on the understanding of the interaction between cells and growth factors or hormones.

They argue that the survival of cells in culture is dependent on the ability of media components to perform the functions previously provided by the normal environment of the cell in vivo. Therefore, the design of defined media should be based on an attempt to reconstruct the extracellular environment involved in supporting cell growth in vivo. This environment involves a combination of hormones, growth factors, binding proteins, and attachment factors. It is argued that although serum supplements may support some cell growth, serum may be an unsuitable substitute for the in vivo environment in many cases. Serum contains substances that may never come in contact with some cells in vivo and the toxicity of high levels of serum has been reported in many instances. This includes reports of the presence in serum of inhibitors of neural differentiation.

Thus, the Sato approach involves the formulation of chemically defined media for optimization of cell growth from a list of isolated and purified growth-promoting substances. These formulations are not perceived as substitutes from serum supplementation but rather as an attempt to simulate the in vivo environment.

Top-Down vs. Bottom-Up Approaches to Serum-Free Media Development

There are two basic approaches to developing serum-free media formulations. The first (top-down) involves taking an existing formulation for a similar cell line, supplemented with serum, and selecting constituents that are stimulatory for growth. This is followed by the slow reduction of serum in the medium. The second approach (bottom-up) involves the selection of a basal medium, analyzing individual components for their effects on growth, and combining them to yield a serum-free formulation.

The top-down approach is often easier to pursue since a working serum-free formulation can often be developed more quickly. Cell lines that belong to the same group, such as epithelial or transformed, often require the same growth factors for growth. Therefore, a formulation that works for one epithelial cell line may work for another with minimal modifications to certain growth factors or hormones (such as EGF, FGF, etc.). For this reason serum-free formulations can be designed faster by this approach.

However, the drawback to the top-down approach is that many components in the formulation may be unnecessary, and often inhibitory for growth. This can often result in the "capping" of the optimal performance of the medium (i.e., the maximum growth may not be achieved) as improvements are hindered by the presence of unwanted compounds.

The bottom-up approach, although more labor-intensive and time-consuming, can lead to higher-quality media. Only the components that are required for growth are included in the formulation, allowing for greater control in optimizing the medium. Thus, media developed in this way tend to have higher growth rates and are more easily improved since inhibitory compounds are less likely to be present.

Plackett-Burman Statistical Approach to Serum-Free Media Development

Another method of serum-free media development involves the use of a factorial experimental design, modeled after the Plackett-Burman statistical approach. This statistical method enables the systematic study of complex components and identifies nutritionally important factors for cell proliferation. In addition, the specific interactions between components (whether beneficial or detrimental) can be observed since the combination of nutrients is more important than the individual components by themselves. A large number of components can be studied at once to determine which factors, or combination of factors, are important for a serum-free formulation. This approach is not limited to cell growth studies and can also be valuable in determining which components in the medium are important for productivity, such as antibody production.

The steps involved in the approach are outlined as follows:

a. Select the basal medium and components that are to be analyzed for their ability to enhance growth or productivity.
b. Establish two concentrations for each component: a "high" concentration and a "low" concentration.
c. Grow the cells in the selected basal medium (e.g., DMEM) with serum (5-10%) in a well-plate culture. Allow the cells to initiate growth and division (~24 hr).
d. Remove the serum-medium and replace with the same basal medium plus combinations of media components (no serum), e.g., DMEM+insulin (high/low concentration)+transferrin (high/low concentration).

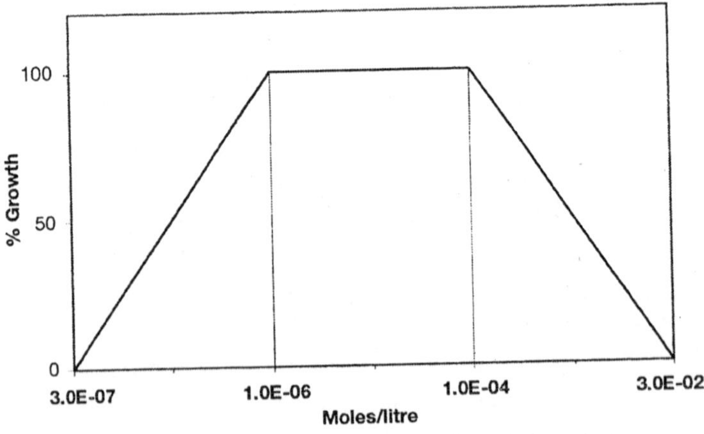

Fig. 3.1 Idealized growth response curve for a single nutrient.

e. Include a positive control (serum-supplemented medium) and a negative control (medium with no components).
f. Count the viable cells (via MTT assay or trypan-blue exclusion method) in each well after –48 hr and/or assay for protein production (productivity).
g. Construct a matrix and determine which combinations of components, and at which concentration, are suitable for encouraging or sustaining cell growth.
h. Calculate the variances from all of the effects from the single factors and their interactions to determine which are significant. This will allow the determination of the best combination of components and thus the most suitable serum-free formulation.
i. Initiate an adaptation procedure for the cells to the new serum-free medium.

This stepwise approach has been utilized in many cases to design serum-free formulations. It has been attempted with the CHO cell line, where a serum-free medium was optimized for human interferon-gamma production. A series of components, including many amino acids, BSA, and other growth factors, were studied. From this, glycine was discovered to affect the specific growth rate, as well as BSA, phenylalanine, and tyrosine. Other amino acids, such as methionine, proline, and histidine played an important role in interferon-gamma production. On the other hand, insulin, arginine, aspartate, and serine produced an inhibitory effect on both cell growth and interferon-gamma production. The end result was an optimized medium that yielded 45% higher productivity than previous formulations.

Similarly, using the same statistical approach, an improved medium for the production of erythropoietin from CHO cells was developed. This medium was based on an IMDM basal medium, supplemented with $Fe(NO_3)_3 \cdot 9H_2O$, $CuCl_2$, $ZnSO_4 \cdot 7H_2O$, insulin, transferrin, and ethanolamine. Glutamate, serine, methionine, phosphatidylcholine, hydrocortisone, and pluronic F68 were identified as positive determinants for cell growth, and the final optimized medium yielded 79% higher productivities than serum-supplemented IMDM.

Future Prospects

The development of better serum-free formulations is ongoing in the scientific and industrial community. The push for the elimination of animal components in the production of biologicals is leading to

TABLE 3.4 PLACKETT–BURMAN MATRIX

	Variables (components)			
Medium	**A**	**B**	**C**	**D**
1	+	+	+	+
2	+	+	+	
3	+	+	−	−
4	+	−	−	−
5	+	+	+	
6	−	+	+	−
7	−	+	−	−
8	+	+		
9	−	−	+	−
10	+			
11	+	+	−	+
12	+	−	+	−
13	+	−	+	
14	+	−	+	+
15	+	−	−	+
16	−			

considerable effort being placed on developing completely chemically defined serum-free formulations. This will probably lead to the elimination of hydrolysates, or other such undefined components in cell culture. The many difficulties associated with serum-free development can make the whole task seem intimidating. The development of each formulation requires considerable investment of time. However, the advantages, such as overcoming regulatory hurdles, increasing the performance of a production system, or eliminating the variability of serum in studies of molecule-cell interactions, can make the investment worthwhile. A greater understanding of cell metabolism, hormone interaction, and nutrient requirements will allow better culture media formulations to be developed in the future.

4

THREE-DIMENSIONAL CULTURE

Three-dimensional (3D) cell cultures have been widely used in biomedical research since the early decades of this century. Holtfreter and later Moscona pioneered the field by their research on morphogenesis using spherical re-aggregated cultures of embryonic or malignant cells. Numerous subsequent *in vitro* studies on organogenesis or expression of malignancy were based on these early investigations. Substantial novel input into research on cell aggregates came from fundamental studies of Sutherland and associates. They pioneered multicellular tumour spheroids (MCTS) as an *in vitro* model for studies on tumour cell response to therapy. MCTS have also been used to study basic biological mechanisms, including the regulation of proliferation, differentiation, cell death, invasion, angiogenesis, and the immune response. This chapter will mainly focus on techniques for growing spheroids, with limited information on 3D cultures other than spheroids.

One major advantage of 3D cell cultures is their well-defined geometry—whether planar or spherical—which makes it possible to directly relate structure to function, and which enables theoretical analyses, for example of diffusion fields. Combining such approaches with molecular analysis has demonstrated that, in comparison to conventional cultures, cells in 3D culture more closely resemble the *in vivo* situation with regard to cell shape and cellular environment, and that shape and environment can determine gene expression and the biological behaviour of the cells. One impressive example is the ectopic implantation of embryonic cells, which can result in malignant transformation, whereas the same cells undergo normal embryogenesis in the uterus. Conversely, terato-carcinoma cells may undergo normal development when implanted into an embryo. One further example is the relative resistance of cancer cells to drugs in 3D culture compared to the same cells grown as conventional mono-layer or in single cell suspension.

MULTICELLULAR TUMOUR SPHEROIDS (MCTS)

MCTS monocultures

The history of MCTS and their relevance to solid tumours

Multicellular tumour spheroids (MCTS) mimic solid tumours better than mono-layer cultures. MCTS have significantly increased our knowledge of radiation response of mammalian cells, intercellular communication, cell invasion, angiogenesis, and neovascularization.

Although attempts to initiate spheroids directly from biopsies for routine individual predictive drug testing had little success, MCTS from established cell lines provide an *in vitro* model to study mechanisms controlling drug penetration, binding, and action.

In terms of three-dimensional growth, MCTS behave like the initial, avascu-lar stages of solid tumours *in vivo,* unvascularized micrometastatic foci, or intercapillary tumour microregions, as shown in *Fig. 4.1.* Growth kinetics can be described by the Gompertz equation

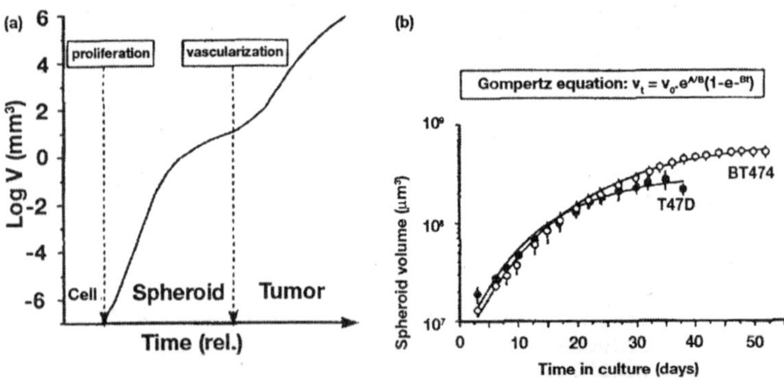

Fig. 4.1. (a) Schematic illustration of the different stages of solid tumour growth that can be mathematically described by the Gompertz function, (b) Representative spheroid volume growth kinetics of two different human breast cancer cell lines grown in liquid-overlay culture in conventional, supplemented media. Lines show non-linear least squares best fits to the Gompertz function with V_0 = 0.498 x 10^7, A = 0.298, and B = 0.060 for BT474 cells, and V_0 = 0.578 x 10^7, A = 0.331, and B = 0.083 for T47D cells.

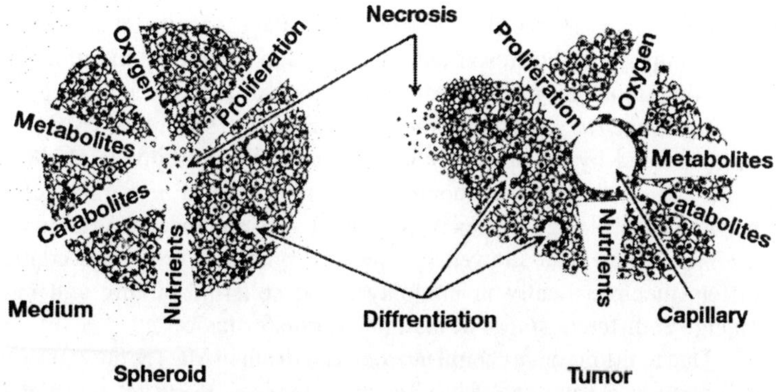

Fig. 4.2 Schematic illustration of the analogy between tumour microregions and multicellular tumour spheroids.

which has been shown to model three-dimensional *in vivo* and *in vitro* growth reasonably well. MCTS develop discrete cell populations similar to those found in microregions of solid tumours with three major groups: actively cycling cells closest to the blood and nutrient supply; quiescent, yet intact and viable cells as intermediates; and necrotic cells/areas furthest from the blood supply (*Fig. 4.2*). Beyond a critical size (500 μm), most spheroids from established cell lines develop massive necrosis in the centre, surrounded by a viable rim of cells with a thickness of 100–300 μm.

Current research with MCTS monocultures

Many of the early findings with the V79 (Chinese hamster lung cells) and the EMT-6 (mouse sarcoma) spheroid models were supported by subsequent studies with human tumour cell lines. NecTotic cell death in the spheroid centre mainly relates to the limited inward and outward diffusion (*Fig. 4.2*). Thus, in some tumour spheroid types, such as WiDr (human colon adenocarcinoma) and Ratl-Tl (*ras*-transfected rat embryo fibroblasts), necrosis and hypoxia are coincident, and ncerotic cell death might be explained by a single limiting factor. In contrast, there is experimental evidence from other MCTS types that necrosis is a complex, multifactorial event.

Investigations on microenvironmental and epigenetic mechanisms involved in the regulation of cell proliferation, differentiation, and gene expression have been intensified enormously within the last ten years. Many cell types seem capable of maintaining intracellular homeostasis in 3D culture until shortly before necrotic cell death, despite environmental stress. There are almost no structural or metabolic markers for this 'pre-necrotic' stage except for a reduced oxygen uptake and/or respiratory activity, and a decrease in mitochondrial function. Cell cycle arrest in the nutrient deprived inner regions of large *myc/ras* co-transformed rat fibroblast spheroids (MR1) is accompanied by an up-regulation of the CDK inhibitor p21kpl/cipl while cell cycle arrest in monolayer cultures and in *myc*-transfected, non-tumorigenic aggregates is associated with enhanced expression of pl8lnk4. These data showed for the first time that cell cycle control differs mechanistically in monolayer and spheroid culture and may change at different stages of malignant transformation.

Due to the massive central necrotic cell death in MCTS, programmed cell death in 3D aggregates has long been ignored. However, recent data show that apoptosis occurs in spheroids of intestinal epithelial cells. Growth inhibition via. stimulation of apoptosis has also been described for p53-defective human lung cancer spheroids transfected *in situ* with a viral wild-type p53 vector. Mueller-Klieser reported apoptotic cell death of multinucleated giant cells in a highly differentiated rhab do my o sarcoma spheroid and in 3D cultures of V79 cells using the TUNEL assay.

3D cultures undergo morphological and functional reorganization. Amongst these activites are:

(a) Modified deposition and/or assembly of bio matrix such as the extracellular matrix molecules fibronectin, collagen types I, III, and IV, or laminin.
(b) Expression and function of ECM receptors (integrins), receptor subsets and subunits, and cadherins.
(c) The expression and function of biological response modifiers such as growth and angiogenic factors (i.e. EGF, TGFcc, VEGF) and/or their receptors.
(d) The development of gap junctional communication.

MCTS culture

A common principle of all spheroid culture methods is the prevention of cellular attachment to a substrate such as surfaces of Petri dishes. Many non-transformed and most malignant cells tend to form spherical aggregates under these conditions.

There are two strategies for cultivating MCTS: in stirred medium and in static medium. In the following section, both these technical

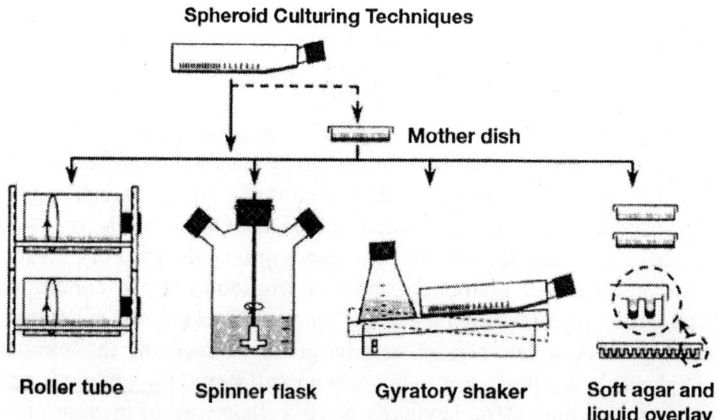

Fig. 4.3. Methods for the cultivatin of multicellular spheroids under continuous stirring (spinner flash, roller tube, gyratory shaker, etc.) or static on non -adherent surfaces (liquid-overlay in 96-well plates).

approaches are described, and the advantages and disadvantages for specific applications are discussed.

The most widely used method for growing MCTS from established cell lines is the spinner flask culture *(Fig. 4.1)*. With this technique a large number of MCTS may be generated simultaneously in large volume cultures, and these can be scaled-up to mass culture. While some tumour cell types are capable of producing cell aggregates directly from single cell suspensions in spinner flasks, cell-cell interaction, and thus aggregate formation, is often more efficient under static conditions. As a consequence, bacteriological Petri dishes or agar/agarose-coated culture dishes, to which cells do not adhere, are often used to initiate spheroid formation.

Cell number and initiation period are critical variables which depend on the cell type and culture conditions (medium, serum pH, etc.). In general, the cell number inoculated varies between 1×10^5 and 2×10^6 exponentially growing cells per 100 mm dish. Once aggregates have emerged in the unstirred medium within two to five days, they can be transferred into spinner flasks. Medium is not refreshed during the initiation period, whereas it is renewed routinely in spinner cultures. Since spheroid initiation may lead to the formation of aggregates with a large variability in size, and since spheroid-to-spheroid clusters may occur, selection of a specific spheroid population prior to the transfer into spinner flask is recommended. If several selective sedimentation steps, for example using Falcon tubes, are not sufficient to eliminate large cell clusters and single cells, hand-selection or rapid harvesting of spheroids by sieving through nylon screens can be performed. The coefficient of variation of the average spheroid diameter within the selected spheroid population should be ≤ 10?

Other frequently used techniques to generate large numbers of spheroids are roller tubes and the 'gyratory shaker' method shown in *Fig. 4.3.* As for spinner flask cultures, an initiation period under non-stirring conditions may be required and the medium should be renewed routinely after transfer into the roller tubes or onto the gyratory shaker. Also, in order to guarantee optimum constant supply conditions throughout spheroid growth, the number of cells per volume of medium should be kept relatively constant by gradually reducing the number of spheroids per flask. For many tumour cell types, the optimum cell number does not extend beyond 3×10^5 cells per ml of medium, and two-thirds of the medium has to be refreshed every 24–48 hours.

The most convenient method of growing spinner flask cultures is in a humidified CO_2 incubator at 37°C, if a stirring device suited for incubator atmospheres is available. Alternatively, spinner flasks may be sealed and kept in a non-humidified, temperature controlled atmosphere or in a thermostatically controlled water-bath positioned over a magnetic stirrer. If the volumes of the gas phase and of the culture medium are approximately the same, the medium of the sealed cultures may be refreshed every 24 hours without generating unfavourable environmental conditions due to cellular production of CO_2 and metabolic waste. The size of commercially available spinner flasks for research purposes ranges from 100 ml to 1 litre with a recommended maximum filling level of approximately 50% of the total volume. The choice of size should be determined by the number of spheroids needed or the (minimum) volume of supplemented medium to be applied. Different types of spinner flasks are available from various companies. While the handling of these spinner flasks, in particular the effort required for cleaning, may differ, the authors have experienced no relevant difference between them in the growth kinetics of MCTS.

An additional parameter that should be optimized is the stirring frequency and thus the shear force in the spinner culture. Optimization should minimize the number of cells that are shed from the spheroids into the medium. Some cell types exhibit an intrinsically high cell shedding in spheroid culture that may not be decreased by the growth conditions. As a result, shed single cells may form small clusters and/or adhere to the vessel wall. In general, these complications can be alleviated, if not eliminated, by changing the culture medium more frequently and by siliconization of the inner surfaces of the spinner flasks.

Some cell types do not form ideally spherical spheroids. Indeed, some rumour ceils do not even exhibit cell aggregation, such as some variants of the MCF-7 breast cancer cell line. If no or only irregular cell clusters are obtained, a number of parameters can be modified to achieve satisfactory cell-cell aggregation and aggregate growth, including:

(a) Initiation of aggregate formation:
- serum type (batch)
- serum content
- culture medium (glucose and glutamine concentrations)
- non-adhesive dish (microbiological, agar-coated)
- CO_2 in incubator (pH in culture medium)
- cell concentration

(b) Initial phase of spinner flask culturing:
- same factors as in (a); each factor may be different from those in (a)
- stirring frequency
- spinner flask geometry

PROTOCOL 1
INITIATION AND CULTIVATION OF TUMOUR SPHEROIDS IN SPINNER FLASKS

- **Equipment and reagents**
- 100 mm agarose-coated culture dishes (*see Protocol 2*) or bacteriological Pern dishes (Falcon)
- 50 ml tubes (e.g. Falcon)
- Different types of pipettes (Pasteur, Eppendorf. 10–25 ml glass or one-way pesof pipettes(Spinner flasks (Techne or Bellco)
- Stirring drive unit
- CO_2 incubator
- Inverted microscope with eyepiece reticule that can be moved with a micrometer screw
- Trypsin/EDTA solution
- MEM, DMEM, RPMI, or other standard medium
- FCS (fetal calf serum)
- Penicillin/streptomycin

Initiation of 'mother dishes'

1. Enzymatically dissociate exponentially growing monolayer cultures.
2. Stop enzymatic reaction by adding appropriate volume of supplemented medium (e.g. DMEM containing 5% or 10% (v/v) FCS).
3. Transfer single cell suspension into Falcon tube and spin down (1000–1500 r.p.m.).
4. Resuspend cell suspension, count cells with automated cell counter or nemo-cytometer. and prepare cell suspension with a concentration of 1×10^5 cells/ml.
5. If the optimum cell concentration to be inoculated per dish is not known, seed 1×10^5, 2×10^5, 5×10^5, and/or 1×10^6 cells/dish each into three or four bacteriological Petri dishes or agar-coated culture dishes with a final volume of 15 ml medium.
6. Grow for four days under standard culture conditions (e.g. 5% (v/v) CO_2 in air in a humidified atmosphere at 37°C). Avoid any motion of dishes—this is important.
7. Check spheroid formation and shape of the aggregates under an inverted microscope and estimate the average size of the round-shaped subpopulation. Spheroids may then be collected and transferred into an one litre spinner flask containing 300–500 ml of supplemented culture medium.

Transfer of aggregates Into spinner flask

1. Collect aggregates from four dishes and transfer into 50 ml medium in Falcon tubes with a Pasteur pipette or an Eppendorf pipette with cut-off tip or a 10 ml glass pipette. The type of pipette may determine the purity of the collection procedure. For example, a smaller tip will not transfer large cell clusters.
2. Let aggregates of appropriate size sediment for 1–3 min and remove 80% of the medium.
3. Add 40 ml of fresh medium.
4. Repeat steps 2 and 3 once or twice to reduce the number of extremely small aggregates and single cells in the suspension.
5. Transfer all or an aliquot of the selected spheroids into a bacteriological or agar-coated dish for accurate sizing.
6. Size a representative population of 25–50 spheroids by measuring two orthogonal diameters for each individual spheroid, either using an inverted microscope equipped with a calibrated reticule

or by applying a calibrated image processing apparatus. Calculate the average spheroid volume.

7. Transfer 50–100% of the selected spheroid population into an one litre spinner flask containing 300–500 ml of supplemented medium. If all spheroids are to be used for further culturing, all steps (1–6) have to be accomplished under sterile conditions.

Cultivation of MCTS in spinner flasks

1. Transfer the one litre spinner flasks into humidified 37°C temperature controlled CO_2 incubator and place onto stirring drive unit with a speed of 75-100 r.p.m. (smaller spinner flasks will need a lower speed).

2. Routinely feed spheroids at day 2 and every first or second day thereafter by allowing spheroids to sediment in the spinner flask for a few minutes and by replacing two-thirds of the medium.

3. In parallel, size a representative population of 25–50 spheroids according to part B, step 6.

4. Estimate the number of spheroids in the spinner flask. Dissociate 10–50 spheroids using a 5 × concentrated enzymatic solution (5 × compared to that used for monolayer cultures of the same cell line), in order to determine the number of cells/spheroid and the number of cells per medium volume unit.

5. Reduce the number of spheroids if the cell concentration exceeds 2×10^5 cells/ml or if pH in the medium drops substantially below 7.0.

6. Prior to a final experiment, hand-selection with a Pasteur pipette or an Eppendorf pipette with cut-off tip or rapid harvesting of spheroids of approximately the same size by pipetting them through a nylon screen may be required.

PROTOCOL 2

COATING OF 96-WELL PLATES OR CULTURE DISHES WITH AGAROSE

- **Equipment and reagents**
- 100 mm culture dishes and/or 96-well plates (e.g. Falcon)
- 10-25 ml glass or one-way plastic pipettes and/or multistep pipette with dispenser tips
- High gelling agarose
- MEM, DMEM, RPMI, or other standard medium
- Penicillin/streptomycin
- Distilled H_2O

Method

1. Prepare up to 200 ml (for 96-well plates) and up to 600 ml (for culture dishes) of 1.5% *(wjv)* agarose in serum-free medium (e.g. DMEM).
2. Autoclave agarose solution for at least 20 min.
3. Let agarose solution cool down to approx. 60°C.

Alternative method

1. Prepare 40 ml (for 96-well plates) and 100 ml (for culture dishes) of 3% (w/v) agarose in distilled H_2O.
2. Autoclave agarose solution for at least 20 min.
3. Warm up (to 37?C) 160 or 500 ml of serum-free medium (e.g. DMEM) and dilute freshly autoclaved 3% (w/v) agarose in H_2O with appropriate amount of medium to 0.5%.
4. (a) Use 20 ml or 25 ml glass pipette to spread 10 ml agarose solution into each culture dish and check for complete surface coating; 60 dishes may be prepared at a time.
   (b) Use multistep pipette with appropriate dispenser tip to spread 100 µl agarose in each well of a 96-well plate; coating of 20 plates is feasible.
5. Agarose-coated dishes and 96-well plates may be stored for approx. one week (depending upon the qualify/strength of the agarose gel) in a humidified atmosphere containing 5% (v/v) CO_2 in air at 37°C. For long-term storage (up to two weeks), dishes/plates may be sealed and stored at 4 °C. These dishes/plates should be incubated under standard culture conditions for 30–60 min prior to spheroid initiation.

Spinner culture is preferable to growth of MCTS on stationary, non-adhesive surfaces, if large numbers of spheroids with a maximum size are needed. Spheroids continuously grown on agar/agarose-coated surfaces may not reach their maximum size, and the thickness of the viable cell rim is often relatively small. On the other hand, various modifications for culturing spheroids in so-called agar- or liquid-overlay *(Fig.* 4.3) have been established. The application of 96-well plates is of particular interest, since individual spheroids can be monitored repeatedly and may be manipulated throughout growth. Following an initiation period under static conditions, liquid-overlay cultures may be kept on a gyratory shaker to guarantee optimum nutrient supply. As a prerequisite, 96-well plates are coated with 0.5-1.5% agar/agarose as described in *PTUUKUI 2.* A novel semi-adhesive substrate used for spheroid cultivation takes advantage of the spontaneous detachment of small cell aggregates formed on semi-adhesive substrates such as proteoglycans or positively charged polystyrene.

Protocol 3
Cultivation of liquid-overlay tumour spheroid cultures in 96-well plates

- **Equipment and reagents**
- Agarose-coated 96-well plates (*see Protocol 2*)
- Various pipettes including an 8- or 12- channel Eppendorf pipette
- Inverted microscope with an eyepiece reticule that can be manipulated with, a microscopewithaney screw
- Gyratory shaker (optional)
- 0.25% trypsin 1% EDTA solution
- MEM, DMEM. RPMI, or other standard medium
- FCS
- Penicillin/streptomycin

Method

1. Enzymatically dissociate exponentially growing monolayer cultures (e.g. with 0.25% trypsin/1% EDTA).
2. Stop enzymatic reaction by adding appropriate volume of supplemented medium (e.g. DMEM containing 5% or 1056 (v/v) FCS).
3. Transfer single cell suspension into Falcon tube and spin down at 300 g.
4. Resuspend cell suspension, count cells with automated cell counter or hemo- cytometer, and prepare cell suspension at a concentration of 2×10^4 cells/ml.
5. If the optimum cell concentration to be inoculated per well is not known, seed 5×10^2. 1×10^3, 2×10^3, and/or 4×10^3 cells/well each into 24 agar-coated wells with a final volume of 200 μl medium·
6. Incubate liquid-overlay cultures under standard culture conditions (e.g. 5% (v/v) CO_2 in air in a humidified atmosphere at 37 °C). Avoid any motion of plates for a minimum of three to four days.
7. Check cell aggregation under an inverted microscope at days 3-4 and size a representative population of 12-24 spheroids by measuring two orthogonal diameters for each individual spheroid either with a reticule manipulated by a micrometer screw in the

eyepiece of an inverted microscope or using a calibrated image processing apparatus.

8. Eventually place 96-well plates on gyratory shaker positioned in the CO_2 incubator for further culturing.
9. Feed spheroids routinely every 24-48 h by refreshing 100 μl of the medium using 8- or 12-channel Eppendorf pipettes and size a representative spheroid population until maximum size is reached.

Additional methodological aspects

In addition to the volume of the spheroids, which should be monitored routinely, analysis of the proliferative activity is useful. The proliferation status can be determined autoradiographically by [3H]thymidine labelling (TLI = thymidine labelling index) or immunohistologically utilizing the BrdUrci antibody technique or antibodies detecting specific proliferation-associated antigens such as PCNA or Ki67. Alternatively, spheroids may be dissociated enzymatically and. fixed or unfixed single cell suspensions can be analysed for their cell cycle distribution using flow cvtometry and DNA-specifk fluorochromes such as propidium iodide, 7-AAD (7-aminoactinomycin D), mithramycin, or Hoechst DNA dyes.

Sections of frozen or fixed, paraffin-embedded spheroids can be stained by any conventional labelling technique, or examined by scanning or transmission electron microscopy.

A method for isolating specific cell subpopulations from spheroids is described below in detail. This automated dissociation technique is based on the sequential exposure of the cell aggregates to an enzyme solution. The experimental set-up consists of autoclavable, standard laboratory equipment mounted on a gyratory platform in a 37 °C water-bath as shown in *Fig. 4.4.* The enzyme solution is continuously pumped through a temperature equilibration coil into a 'racetrack' dissociation chamber which is filled with 15 ml of enzyme solution prior to the dissociation procedure. The chamber consists of two compartments which are separated by a nylon screen with a pore size of 75 μm. Pumping enzyme solution into the outer compartment and out of the inner part generates a flow of liquid through the screen. Thus, the spheroids are kept in suspension in the outer compartment, while dissociated single cells reach the inner compartment, following the fluid flow. The cell suspension is constantly removed from the inner compartment and collected on ice. Sequential aliquots are sampled from different levels within the spheroids.

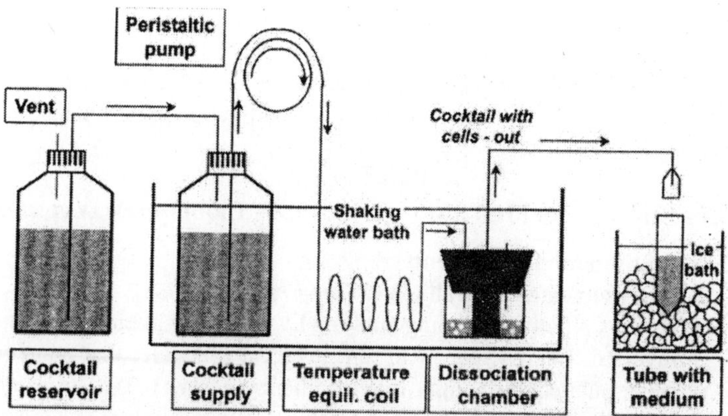

Fig. 4.4 Experimental set-up for automated selective dissociation of multicellular spheroids consisting of an enzyme cocktail reservoir and supply unit, a peristaltic pump for enzyme transport through a temperature equilibration coil into a specially designed dissociation chamber, both placed in a 37°C temperature controlled shaking water-bath. The dissociation chamber contains an outer and an inner chamber which are separated by a 75 μm mesh, and is sealed by a stopper. Spheroids are placed in the outer chamber and the enzyme cocktail is pumped through the chamber so that dissociated cells are transferred into a tube containing ice-cold medium.

MCTS co-cultures

General aspects of tumour heterogeneity

The presence of histologically different cell types within tumours, including host defence cells, fibroblasts and endothelium contributes to their heterogeneity. This heterogeneity can to some extent be mimicked *in vitro* by heterologous spheroids consisting of rumour cells and host cells, Such spheroids may be initiated directly from mixtures of suspended host cells, such as fibroblasts or monocytes and Tumour cells. One disadvantage of this approach is the risk of non-reproducible formation of intraspheroidal clusters of one cell type.

Alternatively, preformed MCTS of defined sizes and histomorphological characteristics can be incubated with suspensions of immune cells such as monocytes or LAK (lymphokine-activated killer) cells, to study immune ceil infiltration and its cytostatic and/or cytotoxic effects on the tumour cells.

Co-cultures of MCTS and immune cells

Mononuclear cells from peripheral blood can be prepared by standard density centrifugation over Ficoll/Hypaque, and human blood monocytes can be separated from this cell mixture by their intrinsic activity to adhere to plastic or glass surfaces. A more convenient technique uses leukaphoresis and countercurrent elutriation, Differentiated macrophages can be generated by long term culture in Teflon bags, and these can be used for co-cultivation experiments.

PROTOCOL 4

CO-CULTIVATION OF MCTS AND MONOCYTES IN LIQUID-OVERLAY CULTURE

- **Equipment and reagents**
- Agarose-coated 96-well plates (*see Protocol 2*)
- Various pipettes including an 8- or 12- channel Eppendorf pipette
- Inverted microscope with an eyepiece reticule that can be manipulated by a microscopewithaney screw 0.25% trypsin/ EDTA solution
- MEM, DMEM. RPM1, or other standard medium
- FCS
- Penicillin/streptomycin

Initiation and cultivation of MCTS and preparation of monocyte suspension

1. Initiate tumour spheroids in agarose-coated 96-well plates as described in Protocol 1.
2. Measure the average spheroid diameter and volume over time, and decide when to add the monocyte suspension (e.g. small size spheroids without necrosis versus rage spheroid with necrotic core).
3. Make fresh preparation of human blood monocytes (e.g. reference 58).
4. Prepare monocyte/macrophage suspension in medium containing 10% FCS (later adding 2–5% AB serum which may be required for macrophage differentiation; the medium and serum should be low in or free of LPS) at a concentration of 1–4 × 105 cells/ml.

Co-cultivation of MCTS and monocytes/macrophages

1. Remove 100 μl of the medium.
2. Add 1–4 × 104 monocytes/macrophages (100 μl of the immune cell suspension) to each well of a 96-well plate containing individual tumour spheroids.
3. Freeze or fix an aliquot of spheroids after 24-48 h of co-cultivation to investigate immune cell migration capacity using

immunohistochemistry and/or dissociate co-cultures to obtain single cell suspensions for further flow cytometric or molecular analyses.

4. Freeze or fix another aliquot of spheroids after six to eight days of co-cultivation to investigate macrophage differentiation using either immunohistochemistry or flow cytometry or molecular analysis following enzymatic dissociation of co-cultures.

Co-cultures of MCTS and fibroblast aggregates

The extracellular matrix (ECM), mainly produced by stromal fibroblasts, has a crucial role in epithelial cell differentiation, growth, and gene expression. In order to investigate interactions between stromal fibroblasts and tumour cells, MCTS can be cultured on confluent fibroblast monolayers or with fibroblast aggregates, as described with human bladder and breast cancer cells. While some tumour cell types overgrow fibroblast aggregates and may show some specific cell-cell interactions in the contact zone, others, in particular highly metastatic tumour cells, seem to invade the fibroblast aggregates.

PROTOCOL 5
CO-CULTIVATION OF MCTS AND FIBROBLAST AGGREGATES IN LIQUID-OVERLAY CULTURE

Equipment and reagents

- Agarose-coated 96-well plates (*see Protocol 2*)
- 100 mm dishes and 50 ml tubes (e.g. Falcon)
- Various pipettes (*see Protocol 3*)
- Sterile scalpel and a pair of scissors
- 0.25% trypsin/1 mM EDTA solution
- DMEM with 4.5 g/litre glucose
- DMEM with 1.0 g/litre glucose
- *FCS*
- Penicillin/streptomycin and amphotericin
- PBS (Mg^{2+}- and Ca^{2+}-free phosphate- buffered saline)
- Acetone (for fixation of monolayers)
- Liquid nitrogen (for freezing)
- 4% (v/v) formalin in PBS for fixation

Preparation of fibroblasts from primary (breast) tumour biopsies

1. Remove adipose and non-tumour tissue from biopsy.
2. Wash tissue with PBS.
3. Incubate tissue for 30–60 min in a high concentration of penicillin/streptomycin (5000 IU/ml penicillin, 5000 μg/ml streptomycin) in PBS.
4. Place tissue in medium and slice into 1-4 mm^3 pieces under sterile conditions.
5. Transfer 20-30 tumour fragments into one 25 cm^2 culture flask.
6. Incubate for approx, 1 h until fragments have adhered to the surface of the culture flask.
7. Cover fragments with DMEM containing 20% (v/v) FCS, 25 mM glucose, penicillin/streptomycin, and amphotericin.
8. Monitor cell outgrowth after two weeks and routinely thereafter.
9. Transfer cells into 75 cm^2 culture flask after sufficient outgrowth using trypsin and reduce serum to 10% (v/v) and glucose concentrations to 5 Mm.
10. Routinely prepare slides with low passage monolayer cultures for fixation and immunohistochemical detection of contaminating cells such as tumour or endo-thelial cells.

Initiation of MCTS and flbroblast aggregates

1. Initiate MCTS and fibroblast aggregates separately according to the routine initiation protocol; if possible use same medium. In general, cell numbers to be inoculated differ for tumour cells (5×10^2 to 2×10^3/well) and fibroblasts (3×10^3 to 6×10^3) by a factor of two to four in order to obtain aggregates within the same size range after an initiation interval of three to five days.
2. Routinely size MCTS and fibroblast aggregates after three to four days and daily thereafter until transfer.

Co-cultivation of MCTS and fibroblast aggregates

1. When tumour spheroids are slightly larger than fibroblast aggregates (e.g. 350 μm versus 300 μm), transfer individual 3D fibroblast cultures into agarose-coated wells, each containing one spheroid, using a Pasteur pipette or an Eppendorf pipette with a cut-off tip (8- or 12-channel Eppendorf pipette may be applied if cut-off tips are available).

2. Monitor growth behaviour every day and feed co-cultures routinely, until fibroblast aggregate is overgrown.
3. Freeze or fix co-culture at different growth stages for immunohistochemical analysis and histomorphological observation of cell-cell interactions. To avoid cell death, do not culture fibroblast aggregates for more than 16 days.

Co-cultures of MCTS and endothelial cells

MCTS are also used in studies of angiogenesis and invasion, such as the embryonic chick heart fragment confrontation culture system in semi-solid medium. Culture of MCTS on confluent endothelial cells with underlying ECM can be used to study tumour cell invasion, but not angiogenesis. Data obtained with melanoma cell spheroids have led to the hypothesis that free radical-mediated endothelial cell damage is one mechanism contributing to melanoma metastasis.

Protocol 6
Co-cultivation of MCTS and endothelial celts

- **Equipment and reagents**
- 100 mm bacteriological Petri dishes and culture dishes (e.g. Falcon)
- Sterile scalpel, pair of scissors, and cannulas
- Inverted phase-contrast microscope
- 70% (v/v) ethanol
- PBS
- Ham's F12/Iscove's medium (1:1)
- FCS

Initiation and cultivation of MCTS

1. Initiate MCTS according to *Protocols 1* and *3*.
2. Routinely feed spheroids.
3. Size spheroids after a defined initiation interval.

Preparation of umbilical cord vein

1. Wash surface of freshly isolated umbilical cord vein with 70% (v/v) ethanol, place into Petri dish, and cover with PBS.
2. Cut umbilical cord longitudinally.
3. Section umbilical cord vertically to obtain 1.5–2 cm pieces and check for integrity.
4. Transfer pieces into new Petri dish and fix tissue with endothelial cell layer upward, e.g, by using the tips of sterile cannulas.

5. Cover umbilical cord tissue with Ham's F12/Iscove's (1:1) medium containing 10% FCS.
6. Incubate under standard culture conditions (5% (v/v) CO_2 in air, humidified, 37°C).

Co-cultivation of MCTS and umbilical cord vein endothelial cells

1. Place spheroids of an appropriate size (e.g. small MCTS without necrosis versus larger spheroids with necrosis) onto freshly prepared umbilical cord vein using a Pasteur pipette or an Eppendorf pipette with cut-off tip.
2. Systematically monitor migration of tumour cells using an inverted phase-contrast microscope.
3. Fix co-culture at different growth stages for routine immunohistochemical analysis and histomorphological observation.

PROTOCOL 7
CO-CULTIVATION OF MCTS AND ENDOTHELIAL CELLS ON BASAL MEMBRANE

- **Equipment and reagents**
- 100 mm bacteriological Petri dishes, culture dishes. 6-well plates, and 25 cm^2 culture flasks (e.g. Falcon)
- Sterile scalpel, pair of scissors, and cultureflas
- Inverted phase-contrast microscope
- 70% (v/v) ethanol
- PBS
- **Hepes** solution: 50 mM **Hepes** pH 7.5,685 mM NaCl. 20 mM KCl,1% (w/v) glucose
- 0.02% cotlagenase in **Hepes**
- 0.2% (w/v) gelatin in PBS
- DMEM (Ham's F12/Iscove's medium, 1:1)
- FCS
- 5% (w/v) dextran
- 0.2 M NH_4OH

Preparation of basement membrane (BM) (68)

1. Wash freshly isolated cattle eyes with 70% (v/v) ethanol.
2. Isolate cornea under sterile conditions using a pair of scissors and a scalpel.
3. Place cornea in sterile Petri dish and scrape off endothelial cells from the inner surface with a scalpel.

4. Transfer endothelial cells into a 5 cm culture dish and cover with 2–5 ml of DMEM containing 10% (v/v) FCS.
5. Culture cells to confluence under standard culture conditions (5–8% (v/v) CO_2 in air, humidified atmosphere, 37°C).
6. Routinely refresh medium every 48 h.
7. Transfer confluent endothelial cells into a 25 cm^2 culture flask and subculture for several passages.
8. Seed 1–5 × 105 cells into each well of a 6-well plate and culture until confluence.
9. Feed confluent monolayers with DMEM containing 10% (v/v) FCS and 5% (w/v) dextran at days 0 and 4.
10. After an additional incubation interval of four days remove medium, rinse with PBS, and lyse cells with 0.2 M NH_4OH (1–2 ml for 10 min).
11. Remove cell debris with PBS, seal 6-well plates, and store at 4°C until use.

Isolation and cultivation of human umbilical vein endothelial cells (HUVEC)

1. Wash surface of umbilical cord with 70% (v/v) ethanol.
2. Rinse the inside of the umbilical cord with 10–20 ml **Hepes** solution.
3. Inject 4 ml of a 0.02% (w/v) collagenase solution in **Hepes** into the umbilical cord vein and carefully seal the edges.
4. Incubate in 37°C water-bath for 10 min.
5. Gently massage the umbilical cord.
6. Open both ends and collect endotheiial cell suspension into 50 ml Falcon tube by injection of 10 ml Hepes solution.
7. Spin down cells (5 min, 300 g) and resuspend in 5 ml Ham's F12/ Iscove's (1:1) medium containing 10% (v/v) FCS.
8. Inoculate cells into a 25 cm^2 culture dish coated with 0.2% (w/v) gelatin in PBS.
9. Incubate cells for 30–60 min under standard culture conditions, monitor cell adherence, wash carefully, and refresh medium.
10. Feed cells every other day and subculture at confluence using 0.02% (w/v) collagen-ase in Hepes.

Co-culttvation of MCTS and HUVEC on BM

1. Inoculate 1–5 × 105 HUVEC onto each well of a BM-coated 6-well plate and culture until confluence using serum-supplemented Ham's F12/Iscove's (1:1) medium.

2. Initiate and culture MCTS according to the routine initiation protocol (see *Protocols 1* and *3*).
3. Transfer individual spheroids with defined sizes onto confluent HUVEC monolayers (two spheroids per well).
4. Systematically monitor migration of tumour cells using an inverted phase-contrast microscope.

Additional technical aspects

Separation of cells from defined locations within sections from 3D cultures can be achieved using microdissection techniques, such as laser microbeam combined with an optical tweezer. Dissociation of co-cultures by enzymatic or mechanical means allows for flow cytometric analyses and for separation of cellular subpopulations by various sorting techniques. As a prerequisite for cell detection and separation, cell sub-populations can be permanently labelled by the introduction of specific marker genes, for example using ?-galactosidase or green fluorescent protein (GFP) or labels for cell membranes. GFP can be detected in routine flow cytometry.

Experimental tissue modeling

3.1 Current research on tissue modelling

Substantial progress has been made towards the development of an artificial liver. Strategies include culture of hepatocyte spheroids or of hetero-spheroids consisting of hepatocytes and fibroblasts or parenchymal cells. The application of artificial support systems, such as porous gelatin sponges, agarose, or collagen, as well as the induction of aggregation by exogenous adhesion molecules is advantageous for long-term culture of liver cells. Reconstitution of the specific geometry, microenvironment, and metabolism of the hepatic sinusoid in a complex 3D culture system has been reported and pig hepatocytes have been grown in co-culture in an artificial 3D capillary system.

Development of an artificial pancreas has focused on spheroids from insulin-secreting recombinant mouse pituitary AtT 20 cells and mouse insulinoma beta TC3 cells. Long-term viability and function of human fetal islet-like cell clusters implanted under the kidney capsule of athymic mice have been achieved. Another promising strategy is the use of isolated piscine islets, so-called Brockmann bodies, as 'natural' pancreatic spheroids which can be implanted as xenografts after microencapsulation with alginate. Oxygen supply seems to play a crucial role for the survival and maintenance of function of such islets.

The pituitary gland has been modelled using multicellular spheroids to study hormone release, including luteinising hormone (LH), following stimulation with the releasing hormone LHRH. Three-dimensional cultures of mela-tonin secreting cells from the pineal gland have also been described. In contrast to pineal cell monolayers, 3D cultures remain functional for more than three weeks secreting melatonin when challenged with isoproterenol. Despite their tendency to spontaneously form follicles in aggregation culture, there are few reports on isolated adult thyroid cells. Thyroid cell spheroids allow study of cell rnotility, cell adhesion, and E-cadherin expression in thyroid follicle biogenesis.

3D brain cultures has been reviewed comprehensively and have been used to study neural myelination and demyelination, neuronal degeneration, and the neurotoxicity of lead. Neuronal behaviour following HIV infection, Alzheimer's disease, and Morbus Parkinson has been studied using aggregated brain cell cultures. Retinal development has also been modelled.

3D chondrocyte research has focused on molecular aspects of matrix formation and differentiation.

Aggregated cell cultures of heart cells have been used to study aspects of cardiac development and physiology, including sinoatrial node cell preparations, atrial cell preparations, embryonic chick heart cell aggregates, and ventricular cell aggregates.

Other tissues that have been modelled in 3D cultures include mesangial cells, urothelial cells, and nasopharyngeal cells.

Tissue modelling of skin and mucosa

Heterotypic 3D cultures including fibroblasts and keratinocytes have been used for studies on the formation of skin and hair follicles, and the influence of mesenchyme on the differentiation of keratinocytes. A method is described in *Protocol 8* and shown diagrammatically in *Fig. 4.5.*

PROTOCOL 8

ISOLATION AND CULTIVATION OF DIFFERENT CELL POPULATIONS FROM ORAL MUCOSA

- Equipment and reagents
- Standard cell culture equipment (including laminar flow hood, incubator, centrifuge, etc.)
- Tissue strainer (70 μm pore size; Falcon)

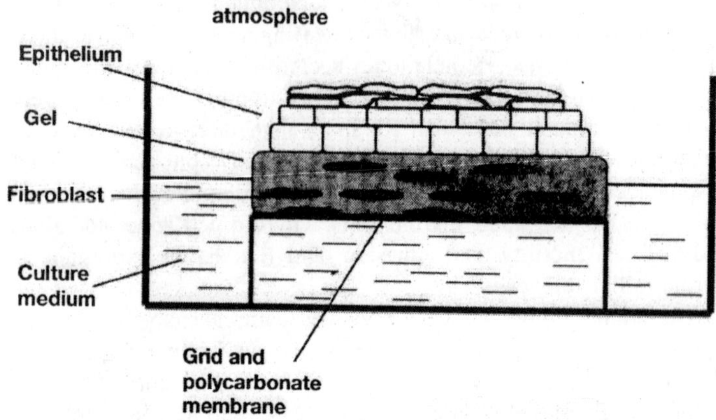

Fig. 4.5 Tissue culture chamber for cultivation of oral mucosa cells. The epithelial cells are exposed to the incubator atmosphere, and may not grow if covered by culture

- Nudeopore track-etch membrane (polycarbonate membrane; Corning Costar)
- Microsurgical scissors and tweezers
- 100 ml dispase grade II (240 U; Boehringer Mannheim)
- DMEM with 4.5 g/litre glucose
- FCS
- Penicillin/streptomycin
- Trypsin/EDTA solution
- PBS (e.g. Sigma-Aldrich.
- Collagen type I
- 1 M NaOH, sterile
- FAD_{spec} medium: 1:3 (v/v) of Ham's F12 and DMEM supplemented with 18.2 μg/ml adenine. 8.33 μg/ml cholera toxin, 10 ng/ml EGF, 0.4 μg/ml hydrocortisone, 5 μg/ml insulin

Isolation of cells

1. Wash tissue in 70% (v/v) alcohol.
2. Rinse tissue in sterile phosphate-buffered saline (PBS). Repeat steps 1 and 2.
3. Separate connective tissue (pink) from epithelium (white) with tweezers, scalpel, and micro-scissors.
4. Mince epithelium with micro-scissors to homogeneity.
5. Mix epithelium with 3 ml dispase II and transfer to a sterile Petri dish.

6. Incubate for 30 min at 37°C.
7. Inactivate enzyme by adding 30 ml DMEM with 10% fetal calf serum (FCS).
8. Spin down for 10 min at 1200 r.p.m. (300 g) at room temperature.
9. Resuspend pellet in 1.2 ml FAD_{spec} medium with 5% (v/v) FCS.
10. Transfer suspension to small (25 cm²) tissue culture flasks and incubate under standard conditions.
11. Mince conneotive tissue with micro-scissors and suspend in l ml DMEM (4.5 g/litrc glucose) with 10% FCS.
12. Transfer suspension to small size tissue culture flasks and incubate under standard conditions.

Cell culture: keratinocytes

1. Incubate epithelial cell suspension at 37°C in a humidified atmosphere containing air and 5% (v/v) CO_2.
2. Replenish FAD_{spec} medium with 5% (v/v) FCS twice a week until cellular outgrowth from adhered cellular islets can be observed.
3. Continue with three complete renewals of the medium.
4. At near confluence, trypsinize cells and pass the suspension through a cell strainer to remove cell clusters.

Flbroblasts

1. Incubate suspension of cells and clusters from connective tissue at 37°C in a humidified atmosphere containing air and 5% (v/v) CO_2.
2. Replenish medium (DMEM, 4.5 g/litre glucose. 10% FCS) on a weekly schedule, until cellular outgrowth from adhered cellular islets can be observed.
3. Continue with replenishment of the medium two to three times per week thereafter.
4. At near confluence, trypsinize cells and remove clusters, as described for the keratinocytes.

Organotypic co-culture

Preparation of fibroblasts

1. Trypsinize fibroblasts and resuspend in FAD_{spec} or DMEM.
2. Determine cell number.
3. Sediment cells by centrifugation (10 min, 300 g, room temperature).

4. Resuspend pellet in an equal volume of FCS.

Preparation of collagen gel

All work to be done on ice near 0°C, unless stated otherwise.

1. Mix 1:10 (v/v) 80% (v/v) collagen type I (4 mg/ml) and 10% (v/v) Hank's buffered salt solution.
2. Neutralize this mixture with 5 M NaOH (to achieve pink colour).
3. Add FCS containing cells to the gel mixture and adjust cell concentration to 1.5–3 × 10^5 fibroblasts/ml gel and FCS content to 10% (v/v).
4. Fill gel suspension into wells of a multititre plate (24-well, 1 ml).
5. Let the gel solidify for 1–2 h at 37°C, and cover it with FAD_{spec} medium.
6. During incubation, fibroblasts should change their shape from a spherical to an elongated form within two days.

Preparation of the keratinocytes

1 Trypsinize keratinocytes and suspend in FAD_{spec} medium.

2 Determine cell number.

3. Overlay 1.5–3 × 10^5 keratinocytes in a drop of medium on top of the fibroblast gel.

Cultivation of the co-cultures

1. Incubate co-cultures for 24-48 h (humidified, 37 °C, 5% CO_2 and air).
2. Replenish medium (careful suction) two to three times per week.

Embryoid bodies

Embryoicl bodies (EB) are derived from embryonic stem cell (ES) cell lines which have retained their capacity to generate cells of the haematopoietic, endothelial, muscle, and neuronal lineages. Development of skeletal muscle myocytes in embryoid bodies is similar to that *in vivo* and beating EB with cardio-specific receptors, ionic channels, and action potentials have been produced. Oxygen-regulated gene expression during embryonic development has also been studied.

5

MECHANICAL SEPARATION

This chapter focuses on the different approaches for separating mammalian cells from cell culture medium containing recombinant proteins of therapeutic interest and their further primary isolation. Today the most common products from mammalian cells are secreted proteins. The isolation of the target protein requires in most cases the initial removal of particles of different physical and chemical properties such as cells and cell debris. This clarification step describes the transition from product generation to product isolation. The unit operation has to perform a proper solid-liquid phase separation of the cell suspension in order to enable further product purification, as the subsequent steps in general require feed streams with low particle occurrence. Upon the removal of larger particles such as cells and debris the clarified cell supernatant undergoes a primary purification step. The main objective of this unit operation is to remove proteins and other molecules with very different physical and chemical character and capture the target protein in a preferably significant reduced liquid volume. If both objectives, capture and volume reduction, cannot be accomplished in one step, an ultra- or diafiltration is performed upon initial capture in order to reduce the liquid volume and transfer the target protein into appropriate buffer conditions for the subsequent fine purification process step. The following paragraphs will give an overview of different methods and established techniques used for the removal of cells and cell debris and the subsequent primary purification of the target protein. Special emphasis is put on techniques developed and applied to clinical and commercial manufacturing of recombinant proteins and patents related to those operations.

Removal of Cells and Cell Debris

The majority of commercially manufactured recombinant products are secreted proteins and the separation of cells or cell debris from the product of interest is an essential process step. However, some applications such as virus production schemes require the harvest of intact cells, as the product remains in most cases intracellular during its expression. In other cases the protein might be associated with the cell membrane like for membrane-anchored receptors. Given the fragility and high value of cell-derived products, cell separation must be accomplished with minimum cell damage and loss of product activity. Choosing the optimal clarification process can be challenging even for microbial-derived products. Although the cell mass in mammalian cell systems is lower than in yeast or microbial-based production processes, a considerable amount of solids has to be removed when operating at a commercial manufacturing scale. The following paragraphs will present in more detail the most common techniques used for separating mammalian cells.

Depth Filtration

Depth filtration is the preferred normal-flow filtration method for removal of cells as the anisotropic character of the filter unit allows a significant higher solid holding capacity than isotropic membranes used for sterile filtration of liquids. Some filters used for clarification combine anisotropic character, large area and capacity with a second layer of isotropic character, which allows sterile filtration quality of the filtrate. Depth filters are no absolute filters. Their highly porous structure allows only a filter rating over a wide range of particle sizes and it enables gas to pass through the matrix. Further details on the manufacturing and structure of depth filters (Fig. 5.1) can be found in Singhvi et al. and several patents from Cook et al. In contrast to other cell separation methods, depth filtration is applied only if the product of interest is secreted and can pass the filtration membrane, as cells cannot be recovered from the filter matrix. Depth filtration systems generally consist of a series of filter cartridges with decreasing pore size rating, thus protecting and extending the life of the following filter

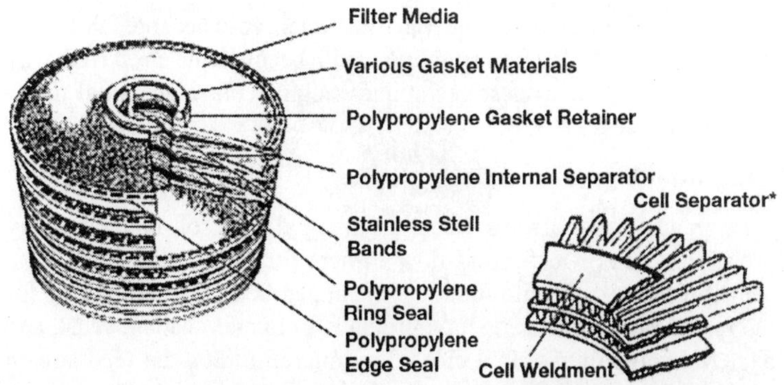

Fig. 5.1 Structure of depth filter. (Courtesy of CUNO Corporation.)

cartridge. As depth filters are not considered sterile filters, the final filter of a depth filtration cascade is usually an absolute 0.2 or 0.1 mm filter. Choosing the correct type and combination of filter cartridge for each application is challenging, but ultimately will lead to an optimized filtration train and reduced overall filtration costs. Depth filters contain diatomaceous earth and are in most cases positively charged. Therefore, separation across depth filtration membranes is based on size exclusion and adsorption. This has to be considered for the correct filter sizing as cells, contaminants, impurities or even product can adsorb to the filter matrix. Filter sizing for depth filters should be undertaken using the volume end point (V_{max}), pressure end point (P_{max}) and turbidity end point (T_{max}) methods. Small-scale studies on CHO cells expressing an IgG antibody showed no loss in product concentration and total quantity upon depth filtration clarification. However, when depth filtration is used in a larger manufacturing scale, a reduced product concentration and some loss of material should be expected due to hold up volumes in the equipment and remaining water from the initial and necessary water rinse of the depth filter. A step yield of about 95% should be expected for process volumes of more than 1000 L. More information on large-scale filtration cartridges can be found in recent patents from Millipore. Depth filtration used for cell removal has several advantages when compared with other methods such as tangential flow filtration (TFF) or centrifugation. In particular, the ease of use, the low initial cost of disposable filter units and the ease of validation allow a rapid and robust development of the clarification step. However, once the manufacturing scale reaches liquid volumes of more than 1000-3000

L, the costs of the disposable depth filter units could become inhibiting. At this scale TFF or centrifugation will become the preferred way for cell removal and these techniques might replace an initial depth filtration process.

Tangential Flow Filtration

Membranes have always been an integral part of biotechnology processes. TFF or cross-flow microfiltration competes with centrifugation, depth filtration and expanded-bed chromatography for the initial harvest of therapeutic products from mammalian, yeast, and bacterial cell cultures. In a cross-flow microfiltration the feed stream is applied in a tangential flow across a separation membrane in order to reduce fouling and clogging of the membrane. The filter unit is incorporated into a

loop system and the feed stream passes the membrane several times. Filtration generally occurs at low pressure with high permeation fluxes across the membrane. As a result, the concentration of particles that cannot pass the membrane increases over time. The filtrate is particle-free and in most cases of 0.2 mm filtered quality thus allowing a subsequent chromatography step without any further particle removal. TFF units are available in various configurations such as hollow fiber,

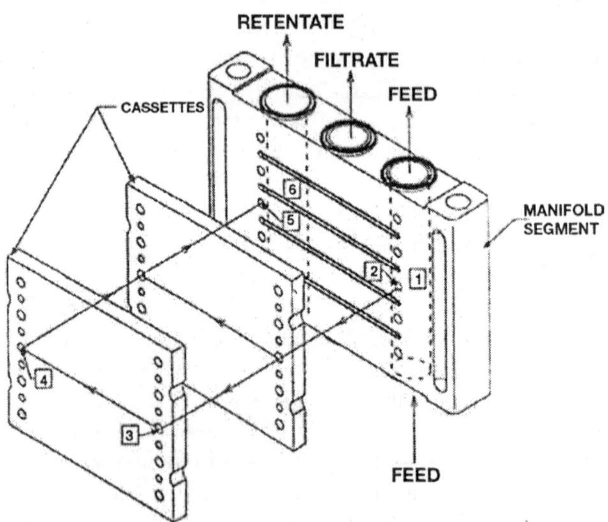

Fig. 5.2 Schematic of plate and frame TFF unit showing liquid flow paths.

spiral wound, flat sheet, tubular, and capillary (Figs. 5.3 and 5.4). A comparison of configuration characteristics and a generic approach to choose the best-suited configuration can be found in a review article from Belfort et al. The most common configuration for cell removal is probably the hollow fiber. Hollow fiber cartridges with larger inner diameter prevent the unit from clogging with cells, however the filtration area per unit decreases with increasing lumen diameter. Robert van Reis, one of the leading authors in this field reported on systems designed for harvest volumes of 12,500 L. The re-circulation rate in the feed stream loop operates at 33,000 L/hr with a filtration rate of 4,800 L/hr. As the filtration process equals a cell concentration generating a cell sludge, the step yield can be improved by washing the cell sludge with additional buffer. Theoretical step yields of more than 99% can be achieved with two diavolumes resulting in a moderate dilution of the feed stream by 7%. Although TFF applies low shear stress to intact cells within the filtration unit, the circulation pump in the feed stream loop becomes a critical and often limiting equipment unit. However, turbulent flow reaching Reynolds numbers of up to 71,000 is reported without causing cell damage. The study concludes that if cell lysis is of no concern filtration can be conducted at high shear rates and in turbulent flow using conditions like those typically applied for bacterial cell concentration. But if cell damage needs to be avoided the tangential flow rate should be limited to an average shear rate of <3000/sec, and transmembrane pressure must be limited to avoid cell deformation into the filter pores.

Tangential flow filtration has several advantages such as potentially 100% cell recovery, a simple product-washing step and a good biological containment. The temperature control is easy and the filtration capacity can be increased rapidly by adding additional modules. The disadvantages on the other hand range from time-dependent permeate flux and long retentate residence times to limited control over separation performance. In some cases the costs for membrane cleaning and replacement could become an additional disadvantage. In order to address some of those disadvantages several improvements were made in recent years. Luque et al. and Gehlert et al. reported on improved filtration performance using helically coiled hollow fiber modules. Chang et al. describe proper selection of operating conditions such as shear rate, flux, membrane pore size, and membrane chemistry. Kwon et al. determined the critical flux at pore sizes ranging from 0.46 to 11.9mm and Li et al. assessed depolarization models. Other authors investigated the mechanism of membrane fouling, impact of fine particles or developed optimization diagrams. Backpulsing has been found as an effective way to reduce membrane fouling and charged membranes or the addition of filtration aids to the feed stream can further improve the clarification performance. Some of the recent patents in this field may serve as a valuable additional literature source. Modifications of the traditional TFF principle use different rotating filter configurations. Vogel and Kroner describe a system with a rotating conical-shaped rotor above a stationary filtration membrane to apply a hydrody-namic lift to cells, which minimizes concentration polarization. Besides the application of cross-flow microfiltration toward cell removal, TFF is widely established for ultrafiltration processes. Observations and improvements made on ultrafiltration processes are described later in this chapter, but can be adapted in many cases to mammalian cell removal processes as well.

Sedimentation

Sedimentation utilizes the density differences between cells or particles and their surrounding medium. Under normal gravitational force the settling velocity of cells or particles in cell culture medium is fairly low. This prevents sedimentation to act as an effective way to remove cells rapidly during a harvest process. Exceptions are processes, which require adherent cells cultivated on microcarriers. The density and mass increase of a confluent microcarrier results in a significantly higher settling velocity and allows for a reasonably short settling and harvest time even at a commercial scale of 2000 L. Gravitational settlers are used only in perfusion processes where a gentle but robust cell separation and retention is required. Various external settling devices are described, but an external inclined settler is the most common settling device in commercial manufacturing. External settling devices are usually cooled to decrease cell metabolism, as the cell retention time in the settling loop can be several hours. The slow operation of settlers is circumvented by increased settling area, thus allowing even perfusion rates of 10 volumes per volume and day. Some novel types of compact settlers have been designed, which allow for a separation between viable and dead cells. These systems take advantage of the higher settling velocity of viable cells over nonviable cells, which shrink upon death. An interesting variation is the sedimentation field-flow fractionation and the sedimentation in a specially designed elongated chamber both applied to blood cells. Overall, the sedimentation technique is most appropriate for selective removal of nonviable cells in longlasting continuous cultivation systems due to the robust nature of the technique and simple equipment. For rapid and quantitative removal of viable and nonviable cells as well as cell debris the settling velocity has to be increased by applying higher gravitational forces to the feed stream. This is the main principle of centrifugation.

Centrifugation

The solid concentrate produced by centrifugation differs from that produced by filtration. At best centrifugation produces a cell paste, but often it yields only a concentrated suspension. Filtration in contrast produces a relatively dry cake thus minimal product loss, which is the major advantage. However, many biological feeds that can be centrifuged cannot be effectively or economically filtered, so that centrifugation is often a very attractive alternative. The five basic types of centrifuges are tubular bowls, disk stack type, basket type, decanter, and disk decanter.

Tubular bowl centrifuges provide very high centrifugal forces resulting in a good dewatering performance. The bowl can be cooled easily, a real advantage in protein work. The liquid feed stream is applied through the bottom and clarified liquid is removed from the top. Solids such as cells deposit on the bowl's wall as a thick paste until the limited solid holding capacity of the bowl is reached. Then the bowl must be dismantled and cleaned, which might be a significant disadvantage for large-scale operations. Tubular bowl centrifuges are usually used to continually separate two liquid streams of different density when the solid content is small. Wang et al. reported on minimal decrease in cell viability using a tubular centrifuge generating g-forces of 19,000 g.

Disk stack centrifuge on the other hand offer continuous operation within a compact volume and the variability in the conical disk stacks allows for large sedimentation areas combined with flexibility depending on the feed streams. The feed usually enters at the top, is accelerated very quickly and the clarified liquid flows out at an annular slid near the feed. Disk stack types can be distinguished between the methods of solid discharge. Solids are either removed intermittently, as in the tubular bowl, involving complete stop of the separation, or solids can be discharged continuously or semi continuously, out of orfices on the side of the centrifuge. The disk-nozzle type allows a true continuous discharge of solids through nozzles in the underflow and liquid through gravity in the underflow. The desludger type on the other hand periodically closes off solid withdrawal, thus causing more compaction and dewatering than the disk-nozzle type. Disk centrifuges have a very high liquid throughput, but if the solids are continuously discharged the liquid content of the discharge is higher than a tubular bowl design, a potential disadvantage if good dewatering performance is required. Continuous disk stack centrifuges are used in perfusion processes or lysis-free cell separation. Takagi et al. found sustained productivity of a tPA-expressing CHO cell when very low g-forces of 67 *g* were applied. Scale-down models are described.

The third type of centrifuge is the basket type, which is essentially a combination of a tubular centrifuge and a filter. It consists of a rapidly rotating perforated basket on which the filter cake accumulates. Liquid passes under centrifugal forces through the cake and perforation and leaves the centrifuge. This type of centrifuge is very efficient for washing accumulated solids such as cells.

The fourth centrifuge type is the decanter, which is also used in truly continuous processes. A decanter can handle a large range of feed concentrations resulting in a dry solid cake. A disadvantage is the decanter's inability to obtain good overflow quality, which means solids are carried into the liquid phase, thus resulting in a less clarified liquid phase. To obtain a dry solid and a clear liquid, a combination of decanter followed by a disk centrifuge is used, which lead to a fifth type of centrifuge, the disk decanter.

The disk decanter is a combination of decanter and disk centrifuge in one single unit. The feed stream first passes through a disk stack and the ejected sludge enters the outer decanter zone where the sludge is further separated to form a dry cake. Although the material strength limitations do not allow for high rotational speeds and therefore only for a limited range of g-forces, the disk decanter can handle large and variable feed concentrations with very good liquid clarity in the overflow.

In addition to the described five main centrifuge types several other types were developed for a special application. A widely used centrifuge for perfusion processes is the Centritech centrifuge from Kendro. The separation chamber consists of a sterile bag stretched over a bowl rotor. Solids accumulate in the lower zone of the bag, whereas the liquid remains in the upper zone. Both zones can be pneumatically separated and the solids and liquids are removed through different outlets of the bag. Sheeler describes a modified fixed angle rotor for continuous-flow cen-trifugation and various patents focus on improving whole blood separation using different bag configurations, separated compartments within the centrifuge bowl, or modifying agents.

The above described cell separation techniques ranging from depth filtration over TFF to sedimentation and centrifugation are all utilized in today's commercial manufacturing processes for recombinant proteins. TFF and centrifugation dominate the commercial scale, where depth filtration can be seen in addition at the clinical manufacturing scale. It is difficult, if not impossible, to describe one of the techniques as the best. Every method has its unique advantages and disadvantages. Choosing the one over the other primarily depends on physical factors such as the cell culture process, shear sensitivity of the host cell, desired clarification grade, or liquid volume to be processed. It often depends also on economical factors such as capital budget, time constraints for delivery, installation and validation, or multi-product considerations. Sometimes a capital budget or ease of operation initially drives a decision in one direction, and later during scale-up for commercial lot sizes another cell separation technique is preferred and needs to be implemented. This change can have a significant impact on the final product quality, which can result in additional work and necessary improvements of the fine purification steps to meet established product quality specifications. A potential change of the cell separation step should be considered and discussed early on in each project.

Product Capture and Primary Purification

Product separation usually begins with the separation of biomass from liquid as described in the previous paragraphs. In many cases, the desired product is in the liquid phase and the separated supernatant will undergo directly further purification steps. The biomass is discarded or sold as by-product.

In some cases, especially in microbial fermentations, the product of interest is not secreted and remains intracellular. Releasing the trapped material involves rupturing the cell wall. Different chemical and mechanical methods can be applied to break the cell wall and release the product. The most common methods are osmotic shock, enzyme digestion, solubilization, lipid dissolution, alkali treatment, homoge-nization, grinding, ultrasonication, and milling. Mechanical techniques are preferred for large-scale applications. As the focus of this chapter is on products from mammalian cells, which are almost all secreted products, the reader interested in microbial fermentations is referred to the literature mentioned at the end of this chapter.

The two objectives of the primary purification step are first to capture the product out of the clarified feed stream with minimal additional binding of by-products or contaminants. The second objective is to significantly reduce the liquid volume for the following fine purification steps, because smaller liquid volumes allow for smaller purification equipment and reduced run times, a major constraint in large-scale operations. To reach both objectives different product capture techniques and methods are used. They range from very economic but less specific methods such as precipitation and extraction to more product-specific but sometimes expensive methods such as affinity chromatography or ion exchange chromatography followed by ultra-and diafiltration. The next few paragraphs will highlight the basic principles and recent developments in those areas.

Precipitation

Protein precipitation describes a process in which the protein of interest is separated from the liquid by adding a reagent to the solution, which in turn

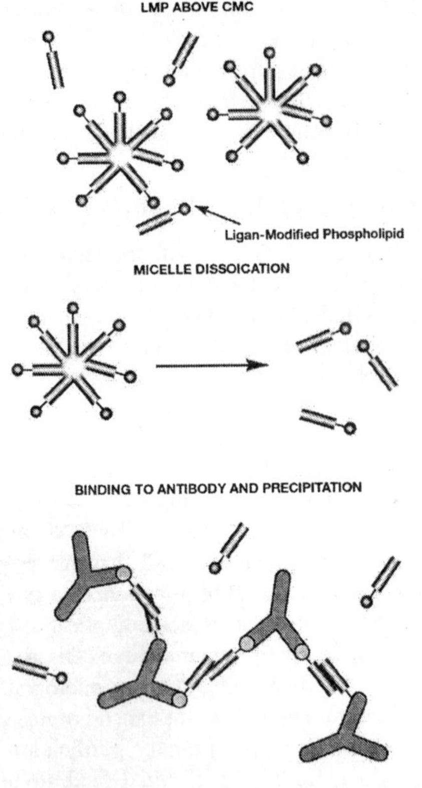

Fig. 5.3 Example of precipitation using ligand-modified phospholipids.

forms an insoluble aggregate with the protein of interests. The insoluble precipitate can be then recovered by centrifugation. Upon recovery, the initially added reagent is removed again. It is the intention to recover the protein in either an unchanged molecular form or one, which can be readily returned to that form. A widely used method for precipitation of proteins is salting-out by salts such as ammonium or sodium sulfate. The salt competes with the protein for water molecules. At high salt concentrations the protein cannot bind enough water molecules to stay in solution and it precipitates out. Although ammonium sulfate is fairly cheap, its corrosive character and difficulty to handle and dispose are sometimes disadvantages for large-scale applications. Another method is the isoelcctric precipitation where the pH of the solution is adjusted to the isoelectric point of the protein of interest. The protein has a net charge of zero and a substantially reduced solubility. The addition of a weak polar solvent such as ethanol or isopropyl alcohol to an aqueous solution of a protein reduces the effective dielectric constant of the solution and the protein solubility decreases considerably. A fourth method is the precipitation by nonionic polymers such as dextrans and polyethylene glycols. The polymer excludes the protein from part of the solution and reduces the effective amount of water available for their solvation. Ionic polyelectrolytes such as alginate or carboxy-methylcellulose act similar to flocculating agents with some salting-out and molecular exclusion action, but they have a high probability of causing structural changes to a protein (Fig. 5.3). Metal ions are used for precipitation because of their high precipitation capability especially at very dilute protein solutions. Metal salts when applied at low concentrations can be removed with subsequent ion exchange resins or through chelating agents. Heat induced precipitation is applied for large-scale plasmid purification. Irwine and Tipton and Gupta reported on a new method using selective agents for affinity precipitation thus overcoming the unspecific nature of protein precipitation. All described precipitation methods focus on creating a well-balanced, reversible, biochemical environment in which the protein of interest has a very low solubility. It is important to evaluate carefully the impact of physical parameters such as pH (Fig. 5.4), ion strength or temperature and engineering parameters such as mixing time, heat distribution, or scalability on the overall process performance when precipitation is considered as protein capture step.

Extraction

Extraction takes advantage of the partitioning of a solute between two liquid phases. The protein is more soluble in one of two liquid phases and the protein concentration increases in the liquid phase of higher solubility as a result of depletion from the other liquid phase. In many cases one phase is water and the second phase is an organic solvent, which is added to the process to initiate the extraction process. Common organic solvents are butanol, amyl acetate, or polyethylene glycol.

However, the low specificity of the method when applied as an initial purification step to a complex protein mixture and the problems related to processing large volumes of organic solvents makes extraction not to a preferred choice for manufacturing processes of cell culture-derived proteins. However, extraction offers a very simple method to transfer a protein out of a very diluted aqueous solution into an organic phase, thus reducing the liquid volume and increasing the protein concentration. The method is used successfully during the initial purification of recombinant proteins derived from transgenic goats and transgenic corn as the starting mixture is less complex than cell culture-derived feed streams (Fig. 5.5). The extraction process can be optimized by using counterions, which binds to the protein of interest, by altering the pH of the aqueous solution or by mixing at low shear rates. Recent developments focused on improving the specificity of the extraction method and elimination of the organic solvent by implementing affinity-based reversed micelles composed of unbound

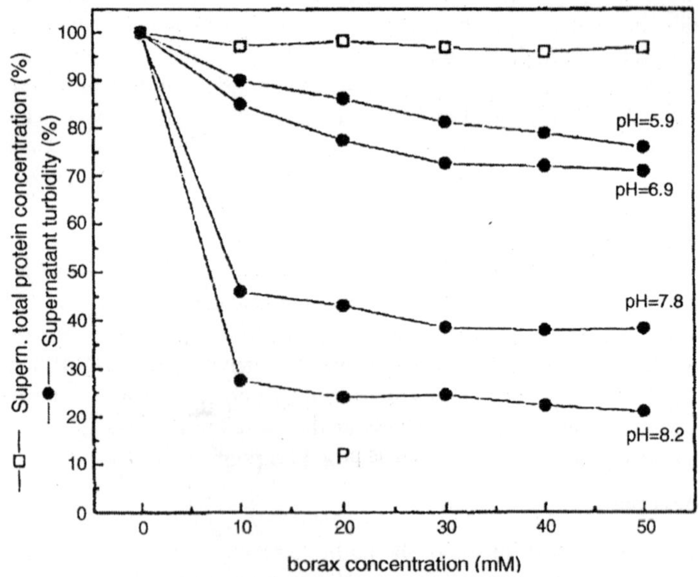

Fig. 5.4 Effect of pH on precipitation by addition of borax.

Cibacron blue in two aqueous liquid phases. These developments address the major concerns the industry has for extraction with organic solvents and they could lead to significant changes in the way commercial product capture is performed today. Performing extraction in a two-phase aqueous system with high protein specificity would allow a direct combination with other purification techniques, which are also aqueous-based systems.

Ion Exchange Chromatography

Ion exchange chromatography is the most used method to capture a protein out of an aqueous solution. The ion exchange chromatography in greater detail and therefore this paragraph will explain the principle only very briefly. Depending on the pI of the protein, either an anion- or a cation-resin is used to bind the protein of interest. Other proteins or contaminants that cannot bind to the resin under the biochemical conditions will pass through the column and are discarded. The column is rinsed with buffer until all nonbound molecules have been removed and the bound protein is eluted with a second buffer. Due to a complex feed stream other proteins or contaminants might bind to the column as well and optimization of the loading and elution conditions is very critical. In order to improve the specificity during the elution step buffer

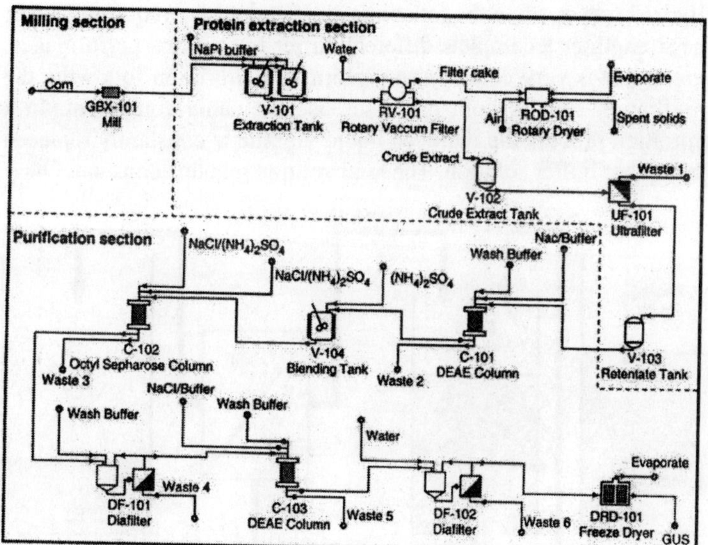

Fig. 5.5 Process flow diagram of purification of recombinant proteinfrom transgenic corn using extraction as first process step.

gradients are usually very shallow or low pH or salt steps are applied. Very often the protein is already 90% pure, but it is more diluted than before in an increased total liquid volume. Then the second objective of any product capture process, a significant volume reduction, can only be accomplished by adding an ultrafiltration step.

Ultrafiltration describes a TFF process. Water can pass the filtration membrane, but the protein of interest cannot, thus increasing the protein concentration and reducing the liquid volume (Fig. 5.6). Choosing the correct membrane grade for a commercial scale is not only dependent upon the molecular weight of the protein, but also upon shear sensitivity, membrane fouling, costs and time constraints, van Reis et al. describe a 400-fold scale-up of an ultrafiltration process at Genentech and large-scale considerations. Several factors influence the performance of an ultrafiltration process and they can be applied easily to any TFF. Burns and Zydney shed light on the effect of the solution pH and Meireles et al. report on effects of operating conditions on membrane fouling. A constant concentration of fully retained protein near the filtration membrane can further improve the filtration performance and product yield as well as charged or modified membranes can help to retain the target protein. The obtained protein concentrate is either frozen and stored until the fine purification resumes or it is directly purified further. However, often the following purification step requires a lower ionic strength or a complete different buffer for optimal performance. Therefore, it is very common to perform a diafiltration following the ultrafiltration step as both processes use the same equipment. In a diafiltration process the removed liquid volume is constantly replaced with another buffer solution. The total volume remains constant. Other

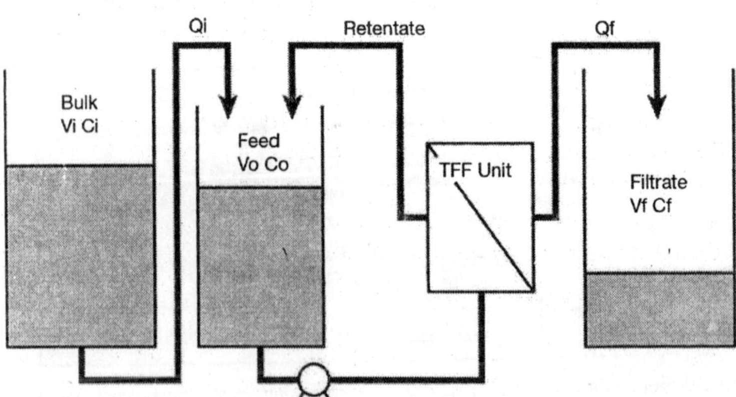

Fig. 5.6 Diagram of tangential flow filtration used for ultra filtration step.

methods such as size exclusion chromatography or counter current dialysis are also applicable for diafiltration, but the TFF is the preferred alternative for commercial manufacturing.

Affinity Chromatography

In order to further improve the capture step toward product specificity and volume reduction, affinity chromatography became very popular over the last 10 years. Affinity chromatography refers to the use of immobilized natural ligands, which specifically interact with the desired protein. The ligand is immobilized on a suitable resin and packed into a column. The protein-containing liquid is passed over the column under conditions under which the protein binds to the ligand. Other proteins or molecules cannot bind to the ligand, pass through, and are discarded. The elution condition, usually a low pH buffer, allows a complete elution of the target protein. Affinity separation is highly specific for the material that binds initially and therefore elution conditions usually do not have to be optimized and tweaked as much as for the previously described ion exchange chromatography, which is less specific for binding properties and requires more optimization of the elution conditions. Very well-defined ligands for the purification of monoclonal antibodies are protein A and protein G (Fig. 5.7), which can be bound to different resins showing similar purification yields, but differences in capacity and pressure drop. A significant disadvantage for large-scale applications of any chromatography step is the limiting flow rate than can be applied to a packed resin. This is especially critical for the initial capture step where large liquid volumes are processed. Two different methods were developed to overcome this problem especially for large-scale applications.

The expanded bed chromatography allows the affinity resin or any other ion exchange resin to expand in the column up to 20-fold of the packed volume. During the loading process the resin is floating in the liquid stream. Upon loading the resin is mechanically compressed, and the protein is eluted into a small volume. Expanded bed chromatography is applied to various production scenarios such as *Escherichia coli*, yeast, or mammalian cells expressing monoclonal antibodies, interleukin-2, interleukin-2 receptor, or interferon. Palsson et al. reported on an improved resin that allows for flow rates as high as 3000 cm/hr.

The second alternative to enable rapid processing of larger liquid volumes is an affinity membrane in a cross-flow configuration. Affinity membranes combine specific adsorption with filtration. Again, protein A and protein G as well as antibo-(a) dies are preferred product-specific

(a)

	Poros 50	Poros LP	Prosep	Sepharose	Stramline
Pressure drop (Pah·cm^{-2})	22.1	12.4	2.1	7.6	0.7
Saturation capacity (g/L)	25	24	26	38	29
Dynamic capacity (g/L)	17.5±0.1	14.8±0.3	13.0±0.3	10.9±0.3	7.5±0.1
Purified antibody					
Yield (%)	104 ±1	106±3	103±2	100±2	105±4
Antibody concentration (g/L)	7.1+0.1	6.4 ±0.4	5.2 ±0.2	3.8 ±0.2	3.2 ±0.1
DNA(ng/mg)	41 ±3	48±3	40±4	29±2	98±18
Host-cell protcins (mg/g)	2.5 ±0.2	2.7 ±0.7	3.7 ±0.2	4.9 ±1.2	6.1 ±20
Protein A (ng/mg)	4.6 ±0.5	7.7 ±0.4	3.1 ±0.5	5.7 ±1.7	6.0 ±1.7
Areal production rate					
R_w (g h^{-1} cm^{-2})	0.17	0.26	0.38	0.20	
U_l(cm·h^{-1})	690	1020	1490	720	
L_c(cm)	10.9	11.3	34.3	28.4	
U_c(cm·h^{-1})	860	1470	1500	750	
Q_s(g·L^{-1})	16.7	10.5	14.6	20.6	
Volumetric production rate					
R_w(g·h^{-1}·L^{-1})	17	23	23	13	
U_l(cm·h^{-1})	700	1000	1000	750	
L_c(cm)	10.3	11.3	11.1	10.6	
L_c(cm·h^{-1})	910	1470	1500	740	
L_s(g·L^{-1})	16.4	10.6	9.5	5.9	

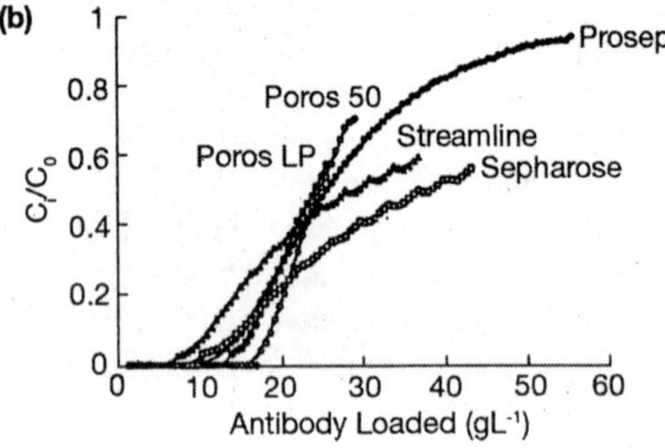

Fig.5.7 Comparison of various resins used for affinity chromatography. (a) Comparative performance values. (b) Breakthrough curves.

ligands that can be bound to a large variety of membrane matrices. Unfortunately, the cleaning of membranes with protein ligands is difficult, as the ligand has to be still active upon cleaning and sanitization of the capture unit. This led to the development of alternative specific ligands that are less sensitive to the cleaning processes in commercial manufacturing. Zeng reviews several membrane types cross-linked with different specific ligands and Tejeda et al. describe a design optimization based on the Thomas kinetic model. Affinity membranes with expensive protein-based ligands are not common in large-scale manufacturing as potential fouling of the membrane combined with the limited or insufficient cleaning regiments would make this process step uneconomical.

6

Advances in Adult Stem Cell Culture

Stem cells are defined as cells that have extensive, some would say indefinite, proliferation potential, differentiate into several cell lineages, and repopulate tissues upon transplantation. Embryonic stem (ES) cells, which are derived from the inner cell mass of a blastocyst, are by all accepted criteria true stem cells capable of generating mature progeny of all cell types. ES cell isolation was initially reported in mouse embryos and more recently in nonhuman primates and humans. When introduced into mouse blastocysts, ES cells can contribute to all tissues of the mouse. Following transplantation in postnatal animals, ES cells generate teratomas, which again demonstrates their pluripotency, but at the same time raises serious concerns when clinical applications of ES cells are considered. Additional legal, ethical, and moral problems accompany the derivation of ES cells from human blastocysts.

Stem cells have also been identified in most adult tissues including bone marrow, brain, skeletal muscle, liver, and pancreas. These findings raise the question of whether adult stem cells preferentially reside within certain tissues or whether they are widely distributed. Stem cells residing in a specific tissue have been traditionally considered as precursors of mature cells of only this tissue, and therefore have been named multipotent, but not pluripotent. Thus, their differentiation potential has been described as limited or tissue-restricted. However, this dogma has recently been challenged and there is growing evidence that adult stem cells may possess a greater plasticity than previously thought.

This chapter is focused on somatic tissue-derived stem cells, their in vitro expansion and differentiation, potential applications as well as concerns associated with recent discoveries and reports related to adult stem cells.

Somatic Stem Cells

A plethora of recent reports has dealt with the identification and isolation of stem cells from various postnatal tissues. Although the majority of papers have described their tissue-specific differentiation potential, in more recent reports, investigators have claimed direct demonstration of tissue-committed stem cells differentiating toward a distant cell type or even de-differentiating to become stem cells for other tissues (transdifferentiation). However, one serious pitfall of many of these studies is that the experiments are rarely performed with single cells. Picking single-stem cells and showing that they can generate progeny of various cell types in vivo or in vitro is the most widely accepted criterion for proving that they are truly multipotent stem cells. True stem cells are clonogenic by definition and this property is the basis for clonality assays proposed to be standard tests when claims regarding transdifferentiation of adult stem cells come into view. Demonstration of clonality should be accompanied by evidence of functional progeny. It is not sufficient to show that the differentiated progeny derived from putative stem cells exhibits certain morphological features or is identified within a specific tissue, e.g., after transplantation. More importantly, the differentiated cells must assume the functional role of the expected cell type. Unfortunately, currently most but a few studies do not provide convincing evidence satisfying these criteria. Therefore, it is prudent—if not essential—to exercise caution before reaching conclusions, as many issues pertinent to adult stem cells are still unsettled. With this in mind, we will mention only a few categories of tissue-specific adult stem cells, as it would be beyond the scope and space limits of this chapter to give an extensive account of studies on all different types of somatic stem cells.

Hematopoietic Stem Cells

Hematopoietic stem cells (HSCs) are probably the best characterized somatic stem cells. Their isolation is based on cell surface markers and functional characteristics. The sialomucin CD34, whose normal function in hematopoiesis remains enigmatic, has become the distinguishing feature for isolation and manipulation of HSCs, although for murine cells that is not necessarily the case. In addition to its expression in stem cells and early progenitors during hematopoiesis, CD34 is also found in vascular endothelial cells and some fibroblasts, necessitating more antigenic determinants for HSC selection. The vascular endothelial receptor 2 (VEGFR2 or KDR), CD133 (AC133), CD90, CD117 and CD38 are some of the markers that have been proved useful in generating functionally homogeneous populations of HSCs. Enriched populations of HSCs have also been obtained via a cell sorting method exploiting the ability of HSCs, that express the BCRP1 membrane pump, to efflux the fluorescent dye Hoechst 33342. This procedure results in a so-called side population (SP), which has recently been identified in bone marrow-derived cells as well as in samples from other tissues.

After transplantation of HSCs into lethally irradiated animals or humans, the neutrophile-macrophage, megakaryote, erythroid and lymphoid cell pool are repo-pulated. In vitro, HSCs can be induced to undergo at least some self-renewing cell divisions and can be induced to differentiate to the same lineages as seen in vivo.

Recent studies have suggested that HSC may also differentiate to nonhemato-poietic cell types. For example, Lagasse et al. demonstrated that bone marrow-derived HSCs were capable to give rise to hepatocytes that could rescue animals with hereditary tyrosinemia. Grant et al. showed that HSC may give rise to endothelial cells that repopulate retinal blood vessels following retinal injury. Other reports have indicated that HSCs can acquire gastrointestinal epithelium, skin,

lung epithelium and muscle as well as neural characteristics. Except for the study by Lagasse et al. and the study by Grant et al., the majority of studies did not demonstrate acquisition of functional characteristics of the distant tissue cells. In addition, aside from the studies by Grant et al. and Krause et al., most studies did not use single cells for transplantation, therefore making it impossible to prove that an HSC acquired the phenotype of a nonhematopoietic cell type. Furthermore, Wagers et al. found levels of transdifferentiation that were significantly lower than what was demonstrated by Krause et al. Therefore, most of these studies only suggest the possibility of "transdifferentiation," without full proof.

Mesenchymal Stem Cells

HSCs are not the only stem cells isolated from bone marrow. Mesenchymal stem cells (MSCs), initially described by Friedenstein et al., are also known to have remarkable differentiation potential as several investigators have reported. MSCs can form osteoblasts, chondrocytes, adipocytes, and skeletal muscle cells. Purification of MSCs is predominantly based on their ability to adhere to culture dishes when incubated with fetal calf serum, but selection with respect to cell surface markers has also been employed for MSC isolation. MSCs are negative for the hematopoietic markers CD34, the leukocyte common antigen CD45, and the lipopo-lysaccharide receptor CD 14. Conversely, cultured MSCs exhibit CD44, CD29, and the transferrin receptor CD71. Furthermore, MSCs express various extracellular matrix components and their receptors such as proteoglycan, collagen (type I, III, IV, V, and VI), laminin, the hyaluronide receptor, and ICAM-1 and -2. Typical marker genes expressed by bone marrow-derived MSCs also include cyto-kines and cytokine receptors such as interleukin (IL)-6 and IL-6 receptor (IL-6R), IL-7 and IL-7R, tumor growth factor (TGF)-b1R and TGF-b2R, basic fibroblast growth factor receptor (bFGFR), and platelet derived growth factor receptor (PDGFR). The expression of cytokines, growth factors, matrix molecules, and receptors is associated with the notion that mesenchymal stem and progenitor cells participate in the organization and function of the stromal microenvironment that promotes differentiation of mesenchymal and hematopoietic cells. Cultured MSCs maintain normal karyotype even after several passages, without differentiating spontaneously.

Depending on the culture conditions, MSCs have been shown to commit toward adipocytes, osteocytes, chondrocytes, skeletal muscle cells, cell types derived from limb-bud mesoderm. In addition, some studies have suggested that MSC may also differentiate into cells with morphological and antigenic characteristics of cardiac muscle cells and neuroectodermal cells. Again, no proof of function has been provided either in vitro or in vivo for differentiation to cells other than limb-bud mesoderm. Donor human MSCs have been detected in liver lung, marrow, spleen, and thymus after transplantation in nonirradiated immunode-ficient (severe combined immunodeficiency, SCID) animals, even though it was not shown that the MSCs acquired morphologic, phenotypic, or functional characteristics of the tissue they engrafted in.

Muscle Cells

Several classes of stem cells exist in skeletal muscle, including muscle satellite cells, and a population of cells that may precede the satellite population. Satellite cells are unipotent myogenic progenitors found in skeletal muscle, characterized by expression of *pax7*, *myf5, C34* and *M-cadherin*, and integrin a-7, that are required for postnatal muscle growth and repair. Satellite cells give rise to large numbers of daughter myoblasts and repopulate the satellite compartment. Recent studies have demonstrated that satellite cells can also differentiate into adipocytes and osteocytes in vitro, indicating mesenchymal differentiation potential of satellite cells. One of the shortcomings of satellite cells for clinical therapy is the limited number of cells that can be harvested, and the low viability following transplantation. In addition, there is no evidence that satellite cells home to and engraft in muscles when administered systemically. Therefore, localized therapies in muscle would be needed.

A second population of cells that, in contrast to satellite cells does not express M-cadherin and can be expanded extensively ex vivo, was described by Qu-Petersen et al. These muscle-derived stem cells (MDSCs) are selected from cultures as cells that can undergo multiple self-renewing cell divisions. MDSC are $CD34^+$, Sca-1^+, and $MyoD^+$, but $CD45^-$, c-kit^-, and M-$Cadherin^-$. Following transplantation in skeletal muscle, MDSC can give rise to long-term persisting grafts of cells that contribute to the satellite cell-like compartment as well as the mature myocyte compartment. In addition, MDSC may also give rise to endothelial cells and neuron-like structures in vivo. Not known is whether these cells will home to muscle when injected systemically.

According to Asakura et al., the SP fraction of skeletal muscle, i.e., the cells that do not label with Hoechst 33342, are progenitors for satellite cells. SP cells in muscle contain both HSCs as well as a population of cells that upon transplantation in skeletal muscle or co-culture with skeletal myoblasts in vitro acquires a muscle fate, including both mononuclear myoblasts and multinucleated myotubes. Mouse SP (mSP) cells capable of differentiation into muscle cells are Sca-1^+, but Myf5-$nlacZ^-$, $desmin^-$ and $Pax7^-$. In contrast to satellite cells, mSP cells are located outside of muscle fibers. When mSP cells are transplanted in skeletal muscle, they can however differentiate into myocytes, and contribute, albeit at low levels, to muscle in *mdx* animals when injected systemically. Jackson et al. published that skeletal muscle SP cells can contribute to the hematopoietic system when transplanted in lethally irradiated mice. Subsequent studies from the same group demonstrated that the hematopoietic potential may be derived from the $CD45^+$ subpopulation of SP cells in muscle. Recently, the $Sca\text{-}1^+$ $CD45^-$ $c\text{-}kit^-$ murine muscle fraction has also been shown to exhibit hematopoietic reconstitution activity following intravenous transplantation into lethally irradiated mice. These putative stem cells derived from muscle could be expanded in vitro without loss of differentiation potential or longevity.

Hepatic Stem Cells

Though normally senescent, hepatocytes exhibit rapid proliferation, driving liver regeneration after partial hepatectomy. This restoration is largely achieved by hepatocyte self-replication leading many groups to suggest that hepatocytes may function as liver stem cells. Others have identified so-called "oval cells" residing within the smallest branches of the intrahepatic biliary network. Oval cells are $Thy1^+$, $c\text{-}kit^+$, and $flt3^+$ and their number increases following hepatocytes loss due to liver disease. The origin of oval cells has long been debated as some groups theorize that oval cells are derived directly from intrahepatic proliferation of cells residing in the biliary tree whereas others support the notion that oval cells have precursors of bone marrow origin. To that end, a number of reports demonstrated that bone marrow hepatopoietic cells could give rise to hepatocytes although this claim has recently been challenged. From an embryo development viewpoint, hepatocytes share their endodermal origin with pancreatic cells and this relation may prove to be important for the generation of hepatocytes from pancreatic tissue cells and vise versa. It has long been known that cells within the pancreas acquire hepatocyte phenotype and morphology in rats placed on a copper-deficient diet. Krakowski et al. found hepatocytes in the islets of Langerhans in insulin promoter-keratinocyte growth factor (KGF) transgenic mice, and similar results have been obtained in vitro following treatment of pancreatic exocrine cells with a combination of dexamethasone and oncostatin M.

Neural Stem Cells

The existence of stem cells in the central nervous system (CNS) was reported in the early 1990s. Employing clonal analysis, labeling and transplantation experiments, rodent neural cell precursors have been shown to differentiate into all three neural cell types: neurons, oligodentrocytes, and astrocytes. Interestingly, neural cell precursors isolated from different regions of the adult CNS and brain, vary in their growth characteristics and differentiation pathways. For example, A2B5 immunor-eactive cells from the rat optic nerve give rise to astrocytes but not oligodentrocytes. Analogously, glial- and oligodentrocyte-restricted precursors have been isolated from other areas. In many instances, neural stem cells are maintained in culture as floating cell aggregates, or neurospheres, in the presence of epidermal growth factor (EGF), but adherent cultures of FGF-dependent neural stem cells have also been reported. In a study by Uchida et al., brain cells that were $CD133^{+}$, $CD24^{lo}$, $CD34^{-}$, $CD45^{-}$, and $5E12^{+}$, were shown to self-renew in neurospheres and differentiate into neurons and glia. Despite the exciting results reported in recent studies on neural stem cells, many issues require scrutiny especially if one considers the use of heterogeneous cell populations in many of these reports.

Multipotent Adult Progenitor Cells

We identified a rare cell within human, mouse, and rat bone marrow MSC cultures that can be expanded for >100 population doublings. This cell differentiates not only into mesenchymal lineage cells but also cells with phenotypic, morphologic, and functional characteristics of endothelium, neuroectoderm, and endoderm. We termed this cell "multipotent adult progenitor cell" (MAPC). Quantitative reverse transcription-polymerase chain reaction (RT-PCR) showed that a transcription factor important in maintaining ES cells undifferentiated, Oct-4, was expressed in MAPC. Furthermore, like ES cells, mouse MAPC, but not human MAPC require leukemia inhibitory factor (LIF) to establish. Mouse MAPC expresses SSEA1 and human MAPC SSEA4. MAPCs are also $Flk\text{-}1^{+}$, $Sca\text{-}1^{+}$, $Thy\text{-}1^{+}$, and $CD13^{+}$, but CD105 and CD44 negative. Single-mouse MAPC injected in a blastocyst contributes to most, if not all, somatic cell types. Upon transplantation in a nonirradiated host, MAPC engraft and differentiate to the hematopoietic lineage, and epithelium of liver, lung, and gut.

Adult Stem Cell Culture

Somatic stem cell proliferation and differentiation in vivo is controlled by the milieu within which these cells reside. Schofield coined the term "niche" to describe tissue regions, which provide a suitable microenvironment to control the fate of HSCs. Such microenvironment can effect both symmetric and asymmetric stem cell division, allowing for normal tissue development and homeostasis while maintaining a pool of cells with extended neogenic potential. Stem cell proliferation and differentiation may be regulated by means of diffuse signaling, cell-cell, and cell-matrix interactions within these regions. The different developmental programs acting in concert during multicellular organization in vivo are mediated by cytokines, tran-scriptional regulators, and adhesion molecules. Many recent reports indicated the instrumental role in multiple differentiation pathways of growth factors such as VEGF, FGF, HGF and their receptors, transcription factors like members of the GATA, Nkx and LIM families and adhesion molecules that include members of the integrin family. Currently, models depicting the action of these factors in determining the stem cell proliferation and commitment are not complete but can serve as a basis for establishing regimens for directional and effective stem cell differentiation in vitro.

Strategies to induce in vitro differentiation of stem cells consist of:

1. *Epigenetic mechanisms:* these take action with the addition of extracellular signals such as growth factors and/or nonphysiological factors (e.g., dimethylsulfox- ide, DMSO) and extracellular matrix components. The presence of these factors aims at modulating the gene expression creating a profile akin to what is observed during developmental processes in vivo.
2. *Genetic mechanisms,* which are mediated by the engineered expression of particular fate-deciding genes.

Epigenetic methods are easier and faster to implement but manipulation of the stem cell phenotype is often incomplete and yields heterogeneous cell populations. Conversely, genetic methods can be more demanding in terms of time and effort but allow for the generation of more pure cell populations. This however may be associated with increased risk for genomic aberrations and malignant transformations. Genetic methods have been applied extensively for directing the differentiation of ES cells. Using for example gene targeting, a drug resistance/toxic gene can be introduced under a cell-type-specific promoter resulting in the selective survival/ablation of a particular group of committed cells. In a similar manner, somatic stem cell differentiation can be directed with the insertion and/or overexpression of a gene that is critical for specification to a certain phenotype. The efficacy of genetic methods for stem cell differentiation depends on a number of factors including how amenable the particular type of stem cells is to genetic manipulation and the targeted gene(s). More information on genetic methods for stem cell differentiation can be found elsewhere and references therein.

Extracellular Signals

Epigenetic methods, in the context of directing stem cell commitment, consist of incubating stem cells with a combination of growth factors and/or nonphysiological agents in order to modulate gene expression in favor of a desired cell type. In most instances, the premise for adding particular growth factors to stem cell cultures for differentiation is that these factors are present in the developing tissue in vivo. The mechanism(s) of action of nonphysiological agents such as DMSO, retinoic acid, and 5-azacytidine in the differentiation of stem cells in the laboratory is even more puzzling. A list of factors used for in vitro differentiation of somatic stem cells is given in Table 6.1. It should be mentioned that studies dealing with in vivo differentiation of adult stem cells (e.g., via transplantation) have not been included in this list.

Besides the factor(s) used for differentiation, other related parameters are also important. These include the factor concentration, time point of stimulus introduction to the culture and duration of cell exposure to the agent(s). For example, different serum concentrations effect skin-derived precursor (SKP) differentiation toward distinct types. In the absence of serum, SKPs acquire a neuron and glial cell phe-notype, whereas adipocytes are generated after incubation with 10% fetal bovine serum (FBS). Moreover, when SKPs differentiate in 3% rat serum they generate a small subpopulation of smooth muscle cells. Like the factor concentration, similarly important for the outcome of the differentiation, are the time point at which a factor is added to the cultured cells and the duration of the treatment. This dependence is expected as many developmental programs for cell commitment and maturation rely on presentation of cytokines to the cells by the extracellular milieu in a temporal manner. Many protocols try to mimic this temporal profile of cytokines. For example, retinoic acid applied to mouse ES cells induces time- and concentration-dependent

TABLE 6.1 FACTORS MEDIATING IN VITRO DIFFERENTIATION OF ADULT STEM CELLS

Adult stem cell differentiation to	Factor(s)
Neural cells (neurons, astrocytes, oligodendrocytes)	Basic FGF (FGF-2), β-mercaptoethanol, butylated hydroxyanisole, EGF, Brain-derived neurotrophic factor, glial-derived neurotrophic factor, PDGF, neural survival factor-1
Endothelial cells	VEGF
Hepatocytes	HGF, FGF-4, oncostatin M, dexamethasone, KGF
Cardiac muscle cells	Co-culture with neonatal cardiomyocytes, 5-azacytidine, TGF-β1
Skeletal muscle cells	5-Azacytidine, amphotericin B, dexamethasone, hydrocortisone, co-culture with myoblasts
Adipocytes	Hydrocortizone, isobutyl-methylxanthine, indomethacin, dexamethasone, insulin, co-culture with adipocytes, serum
Bone and cartilage cells (chondrocytes, osteocytes, etc.)	Dexamethasone, β-glycerolphosphate, TGF-β3, ascorbate-2-phosphate, co-culture with osteoblasts, indomethacin, TGF-β1
Blood cells	Stem cell factor, IL-3, IL-6, erythropoietin, methylcellulose culture

differentiation toward neuronal, cardiac, myogenic, adipogenic, and vascular endothelial smooth muscle cells.

Except from differentiation, various combinations of factors and adhesion molecules have been used to maintain stem cells in a proliferative and undifferen-tiated state in culture. Thus far, the most commonly used factor for maintaining and expanding murine ES cells is the LIF. Similarly, mouse-derived somatic stem cells, for example mMAPCs, require the addition of LIF to the culture medium for expansion. If a murine stem cell-based system is considered, such as for production of a certain metabolite, then the cost arising from the use of LIF can be an important economic parameter. So far, the mechanism(s) by which LIF prevents adult stem cells from differentiating is unclear, but it is postulated that a modus operandi similar to that for ES cells exists. In ES cells, LIF binds to a receptor complex consisting of the LIF receptor and the gpl30 receptor. Upon binding of LIF to the receptor complex, activation of the STAT3 transcription factor ensues. STAT3 may interact with Oct4, another transcription factor whose expression is one of the hallmarks of undifferentiated pluripotent cells. Oct4 expression is evident in murine and human MAPCs albeit in lower levels compared to ES cells.

Oct4 is not the only ES cell marker found in MAPCs or other somatic stem cells. A partial list of stem cell markers is given in Table 6.2. The marker expression pattern depends on a number of factors including the tissue and the species from which the cells were isolated. For example, mMAPCs express stage-specific embryonic antigen 1 (SSEA-1) but no SSEA-3, and the same is true for murine ES cells. A set of stem cell markers should be established for a certain type of tissue-specific stem cells [e.g., nestin has been postulated to be a marker

TABLE 6.2 MARKERS EXPRESSED IN UNDIFFERENTIATED STEM CELLS

Marker	Comments
Oct-3/4	Octamer-specific protein present during mouse embryogenesis (144)
Rex-1	Transcription factor expressed in early embryo (145)
Telomere length	Human: ~11 – 15 kb (65), mouse: ~27 kb (64)
Telomerase activity	Remains high in undifferentiated cells
SSEA-1	Detected in mouse stem cells
SSEA-3 and -4	Detected in human but not mouse stem cells
Sca-1	Member of the Ly-6 family of GPI-linked surface proteins (146)
Thy-1, Thy-1.1, Thy-1.2	Also known as CD90 (147), CD90.1 and CD90.2, respectively

has been postulated to be a marker of brain- and pancreas-residing stem cells]. This can serve as quality control, or more specifically to assess whether the cell population retains its multipotency.

Most reports do not indicate the presence of feeder layers as a requirement for adult stem cell cultures. The presence of feeder layers if needed, would pose a considerable problem for the development of large-scale cultures. Murine somatic stem cells though require the addition of LIF to the culture medium for undifferentiated expansion.

Physical Conditions and Basal Medium

As with any other cell culture system, ex vivo expansion of stem cells in bioreactors requires suitable physiological conditions such as temperature, oxygen and carbon dioxide tension, pH, serum, and medium composition. In addition, some of the above parameters should be adjusted to levels that are essential for in vitro cell growth and depend on the specific stem cell type and its source. Nevertheless, adult stem cells are cultured under the same conditions as most common tissue cell cultures, i.e., at 37°C, 5% CO_2 in medium with pH at 7-7.2. Basal medium formulations may comprise, but not limited to, minimum essential medium (MEM), Dulbecco's modified essential medium (DMEM) and various MCDBs, providing necessary nutrients such as amino acids and salts. Supplementation with insulin, transferrin, and antibiotics is also common.

Serum

Like ES cells, adult stem cells are normally maintained in medium supplemented with certain percentage of serum. The use of serum is associated with a number of potential problems including physiological variability, lot consistency, availability and contamination with growth inhibitors and viruses. These problems can be minimized with the use of high-quality ES cell-screened serum but its high cost may prove prohibitive for large-scale cultures. To that end, low-serum or serum-free media formulations are greatly desirable. We are currently growing MAPCs in 2% fetal calf serum, which is advantageous compared to the 10-15% serum content required for ES cell culture. Furthermore, serum is removed during in vitro differentiation of MAPCs into endothelial, hepatic, and neuronal cells. Therefore, the development of serum-free medium for stem cell culture is feasible.

Cell Density

Another important determinant in a culture system is the cell density and this appears to be especially crucial in the case of stem cells. From our experience with ES cell cultures, we know that once ES cells reach a certain density, they lose their self-renewal capacity and start to differentiate uncontrollably necessitating frequent culture splitting. Differentiation is also prominent when ES cells are cultured on a substratum promoting the formation of aggregates, or embryoid bodies. A similar link between cell density and differentiation appears to exist for tissue-specific stem cells. Densities above 1500-2000 cells/cm favor commitment of cultured mMAPCs to various cell types instead of undifferentiated proliferation. Regardless of whether cell-cell contact or some unknown diffusible signal (or both) may promote differentiation, such low density is an obvious impediment for the use of MAPCs in clinical applications requiring large cell numbers. Higher cell densities have been reported, for example, for cultures of human CNS stem cells. Single cells can initiate the formation of aggregates, or neurospheres, which increase in size as the cells divide. The cultured neurospheres are gently dissociated to a single-cell suspension and passaged every 2-3 weeks on plastic tissue-culture substrata. However, as the neurospheres grow in size, oxygen and nutrient transfer limitations become important, prohibiting in turn higher cell densities. A solution to this problem may be the use of stirred suspension bioreactors. When neurospheres are cultured in suspension, the maximum cell density is improved compared to T-flask cultures, possibly due to the superior mass transfer commonly achieved in suspension bioreactors. The final cell density is boosted even further by increasing the oxygen tension (from 5% to 20%). Upon subculturing, cells retain their capacity to generate new aggregates and to differentiate into all of the primary cell phenotypes in the CNS.

Quality Control in Adult Stem Cell Culture

Quality control assays are essential in adult stem cell cultures as in any other type of cell culture and must be performed frequently and on a regular basis. Of course, quality control assays related to sterility, culture medium, and other parameters are the norm. However, here we would like to focus on assays related to the quality of the cultured stem cells per se. We will limit our discussion on three assays but it is obvious that performing additional assays will help to ensure high culture quality while preventing potential costly and time-consuming problems down the road.

Mycoplasma Testing

Mycoplasma are common contaminants of tissue culture cells and can be transmitted from culture to culture by contaminated tissue culture facilities and/or poor sterility techniques. In fact, mycoplasma contamination appears to be endemic in many laboratories. Part of the reason is that mycoplasma can be latent for long periods making its detection difficult in asymptomatic cultures. Sudden surges of mycoplasma contamination result in changes in cell morphology/ function and eventually cell death. Furthermore, it can be extremely challenging to rescue contaminated cells. Treatment with antibiotics such as gentamycin may be effective against mycoplasma. In general, however, the uncertainty about the effect of the mycoplasma or treatment on the pluripotency of the cells makes as the best solution the immediate disposal of all contaminated cultures and frozen cells. Therefore, it is important to monitor cultures for mycoplasma, for example, once every month. For routine monitoring, nucleic acid-staining dyes may be used allowing for visualization of mycoplasma (disperse punctate cytoplasmic pattern). Other mycoplasma tests with higher sensitivity are based on amplification and detection of mycoplasma-specific ribosomal RNA in tissue culture medium. Along with mycoplasma, provisions should be made for testing possible contamination from other infection agents due to animal products (e.g., serum, animal cells).

Cytogenetic Analysis

By cytogenetic analysis, we refer to chromosome count and karyotype determination. Analysis of the chromosome complement is commonly performed by inducing the cells in metaphase arrest (by incubating with colcemid). Subsequently, the cells are suspended in hypotonic solution causing them to "burst" and spread their nuclear contents. Chromosome spreads are visualized by staining with Giemsa or quinacrine. Ideally, the chromosome number is counted in 30-70 spreads. For kar-yotyping, images are collected from 10 to 15 chromosome spreads and individual chromosomes are sorted into sequence and compared to the expected karyotype of a particular species.

Recently, more sensitive methods have been developed for the detection and analysis of chromosomal imbalances. Comparative genomic hybridization, which is performed on microarray slides, has greatly contributed to the current knowledge of genomic alterations associated with constitutional and sporadic genetic diseases. Furthermore, a novel CGH approach, termed matrix-CGH, where chromosome preparations are substituted by sets of well-defined genomic DNA fragments, exhibits superior resolution and potential for automation.

These methods can be implemented to address putative chromosomal alterations in stem cells and to obtain information about the species and sex from which the cells were derived. This is especially important if adult stem cells from more than one donor or species are simultaneously cultured in the same laboratory. In addition, there is always a possibility that the karyotype may be altered during cell culture. Cytogenetic analysis provides the means to monitor whether the cells retain their original karyotype even after multiple rounds of expansion. The presence of malignant cells can be easily diagnosed, for example by CGH, because their karyotype and/or chromosome number is altered compared to normal cells, and this assessment is imperative if the use of cells is intended for therapeutic applications.

Differentiation Capacity

One of the major characteristics of stem cells is their ability to differentiate toward specific phenotype(s). Cell differentiation protocols, which are established in the laboratory and/or published in the literature, should be employed frequently for testing of the neogenic potential of the cells. In our laboratory, we often test the potential of MAPCs to acquire phenotypes, each of which is associated with a different germ layer. Following are the most common differentiations we perform with MAPCs:

1. Endothelial cells (mesoderm): cells are plated at a density of 20×10^3 cells/cm^2 and treated with 10 ng/mL VEGF. After 10-14 days, cell morphology is akin to endothelial cells and distinct from undifferentiated MAPCs. Furthermore, endothelial markers such as von Willebrand factor, CD62E, CD31, and vascular endothelial (VE)-cadherin can be detected by immunocytochemistry.
2. Hepatocyte-like cells (endoderm): in this case, the plating density is approximately 18×10^3 cells/cm^2 while 20 ng/mL HGF and/or 10 ng/mL FGF4 are added for incubation. As with endothelial cells, cell morphology is changing drastically over the next 2

weeks accompanied by the presence of liver-specific markers such as albumin, cytokeratin 18 (CK18), and α-fetoprotein.

3. Neural cells (ectoderm): incubation of 3–5 × 10^3 MAPCs/cm^2 with 10 ng/mL bFGF results in phenotypes resembling oligodentrocytes, astrocytes, and neurons. The resemblance is morphological and functional as many neural markers are detected including neuron-specific enolase, GABA, Tau, microtubule associated protein (MAP) -2 and serotonin.

The protocols presented here have been applied with modifications for differentiations of other non-MAPC stem cells.

Except from these assays, other methods can be implemented for quality control of the undifferentiated stem cells/committed progeny (e.g., restriction fragment length polymorphism (RFLP) technique can be employed when considering the donor in stem cells of human origin, etc.).

Phenotypic Characterization During In Vitro Differentiation Of Stem Cells

The phenotype in stem cells undergoing in vitro differentiation is commonly assessed by detecting markers specific to a particular cell type. Identification of a single marker does not provide sufficient evidence for determining cell commitment. Normally, cultured cells should exhibit a group of markers related to a certain phenotype (Table 6.3).

Obviously, cell morphology by itself is not sufficient to draw conclusions regarding cell differentiation but it should have a complementary role. It is also a good practice to use multiple techniques such as immunocytochemistry, flow cytometry, RT-PCR and microarrays in order to minimize possible artifacts associated with the implementation of a single method. Determination of the final phenotype should not be limited to the identification of particular markers but should include functional assays such as substrate metabolism by tissue-specific enzymes, electrophysiological measurements etc., depending on the progeny. Lastly, it should be kept in mind that most in vitro differentiations involve starting cell populations which are not totally homogeneous and probably not all the cells will differentiate at the same pace or direction. Therefore, selection and quantitation (e.g., percentage) may be necessary when in vitro stem cell differentiation is assessed.

Large-scale Stem Cell Culture Systems

The rapid progress in stem cell research fueled great hope for effective treatment of a number of pathological conditions. In most studies, the therapeutic potential of stem cells has been demonstrated in animal models, usually rodents, via cell transplantation and subsequent amelioration of conditions such as diabetes, heart infarction, and dopamine deficiency. However, one of the major obstacles for translating such therapies to human patients is the requirement for large numbers of cells. This prerequisite emphasizes the importance of engineering culture systems of sufficiently large scale for ex vivo generation, maintenance, and differentiation of stem cells. A solution to this problem may be the use of bioreactors, which have been used for long time for the generation of large quantities of cells and cellular products. Bioreactors with capacities between 1 and 1000 L may be adequate for the expansion and differentiation of stem cells in clinically relevant numbers, although the size of the cell culture system will eventually depend on the specific end-application as well as on cost-based criteria. The size of the bioreactor is also intimately related to the reactor type. Of course, the development of large-scale bioprocesses assumes solid understanding of aspects of stem cell biology, which are currently the focus of intense investigation. Although a wide variety of bioreactor configurations is available for cell culture, we will limit our discussion to reactor types that have already been employed for the expansion and/or differentiation of stem cells, particularly bone marrow-derived stem cells.

Stirred Suspension Bioreactors

In the laboratory, stem cells are typically cultured in static configurations including multiwell plates, T-flasks, and blood bags equipped with gas-permeable membranes. These static culture systems are not easily amenable to scale-up, on-line monitoring, and control of culture parameters. Conversely, stirred suspension bioreactors are ideal for high-volume cultures of stem cells (Fig. 6.1). These types of culture vessels are readily scalable, simple in their operation, and their technology is mature. In addition, they provide a homogeneous environment allowing for better control of culture conditions. Several groups have tried to expand stem cells, primarily hema-topoietic, in stirred suspension bioreactors. Compared to static cultures, stirred suspension cultures fared better in terms of hematopoietic cell proliferation in the presence of Steel factor and IL-3.

TABLE 6.3 MARKERS ASSOCIATED WITH SPECIFIC CELL PHENOTYPES

Cell type	Markers
Neural cells (astrocytes, neurons, oligodendrocytes)	Nestin, microtubule-associated protein-2 (MAP2), Tau, neurofilament-200[a], neuron-specific enolase, GABA, serotonin, glial acidic fibrillary protein (GFAP)[b], galactocerebrocide (GalC)[c]
Endothelial cells	Vascular endothelial (VE)-cadherin, platelet endothelial cell adhesion molecule-1 (PECAM1, a.k.a. CD31), Flt-1, CD62E, Tek, von Willebrand factor (vWF), vinculin, vascular endothelial growth receptor factor-2 (VEGFR-2)[d]
Hepatocytes	Hepatocyte nuclear factors (HNFs), α-fetoprotein, cytokeratin 18 (CK18), transyretin, transferrin, albumin, urea, dipeptidyl peptidase IV (DPPIV)
Pancreatic cells	Nestin, neogenin3 (ngn3), PDX1, Nkx6.1, Isl-1, insulin, Glut-1, Glut-2
Cardiac muscle cells	Nkx2.5/Csx, α-myosin heavy chain, atrial natriuretic peptide (ANP)[e], a-actinin, cardiac troponin I, desmin
Skeletal muscle cells	Dystrophin, MyoD, myosin, myogenin, MRF4, Myf-5
Adipocytes	Neutral lipid vacuoles[f], peroxisome proliferation-activated receptor γ2 (PPARγ2), lipoprotein lipase (LPL), fatty acid-binding protein αP2
Bone and cartilage cells (chondrocytes, osteocytes, etc.)	Glycosaminoglycans[g], collagen type I and IIh, alkaline phosphatase, calcium accumulation, osteopontin, osteocalcin, osteonectin

[a] Expressed in neurons.
[b] Expressed in astrocytes.
[c] Expressed in oligodendrocytes.
[d] VEGFR-2 is the mouse ortholog of FLK-1 and human KDR.
[e] Also known as atrial natriuretic factor.
[f] Detected by oil red staining.
[g] Detected by toluidine blue staining.
[h] Alizarin Red S stains both collagen type I and II. Antibodies are also available for specific collagen types.

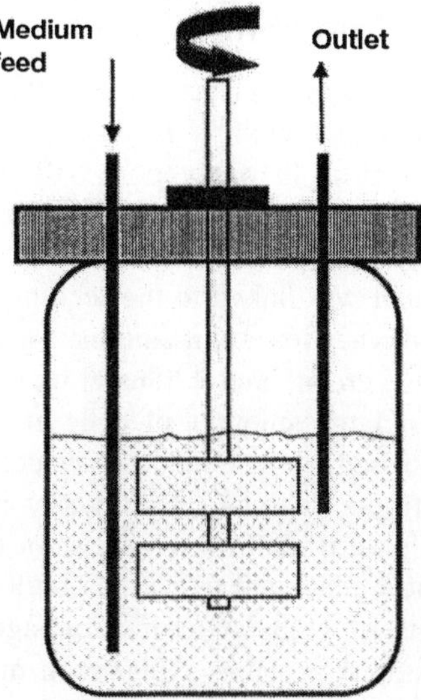

Fig. 6.1 Stirred tank bioreactor.

Suspension bioreactors may be used not only for stem cells expansion but also for evaluating possible mechanisms related to stem cell physiology. In studies regarding the effects of different culture variables on the propagation and differentiation of bone marrow- and umbilical cord blood-derived cells, the rates of progenitor expansion and cytokine depletion were measured and compared among cultures of various bone marrow fractions in the presence of IL-3, IL-6, IL-11, and stem cell factor (SCF). The results suggested the possibility that a feedback mechanism based on cytokine depletion controls the proliferation and commitment of hematopoietic progenitors. In addition to cytokine depletion, agitation, and inoculum density, oxygen uptake and glucose metabolism were shown to affect the rate of blood progenitor cell expansion in stirred suspension bioreactors.

These reactors have also been used for the propagation of neural stem cells in cell aggregates or neurospheres. After growing the cells for over a month, 107-fold higher cell number was obtained without reduction in growth rate and viability. The cells retained a normal karyotype throughout 25 population doublings and were able to differentiate toward neurons, oligodendrocytes, and astrocytes. The effect of agitation and medium viscosity were also investigated in these studies.

Perfusion Chamber Bioreactors

Another bioreactor system that has been studied for large-scale propagation and differentiation of stem cells, primarily hematopoietic, is the perfusion chamber. In one report, 10-30-fold expansion of human bone marrow mononuclear cells (BMMCs) was achieved over 2 weeks of culture starting with inocula of ~10^7 cells. The extent of increase in cell number was linked to the seeding density, the time of cell harvest and the oxygen tension maintained within the system. In a later study, oxygen and growth factor transfer were found to be limiting for the generation of large amounts of cells in perfusion chambers. Compared to unprocessed whole bone marrow cells (WBMC), CD34-enriched cells on stroma exhibited a markedly increased expansion (1300-fold vs. 13-fold for BMMCs) over the same period. Interestingly however, the WBMCs generated five times higher concentrations of colony-forming units of granulocyte-macrophage (CFU-GM/mL of bone marrow aspirate). A possible explanation may be the increased amounts of EGF and platelet-derived growth factor (PDGF) present in the WBMC cultures. As in the case of stirred bioreactors, the effects of cytokine consumption kinetics have been studied in perfusion chamber cultures of hematopoietic cells with the intention of elucidating interactions among pathways regulating stem cell proliferation and differentiation.

Furthermore, the geometry and transport phenomena in various perfusion bioreactor configurations have been studied in order to increase the efficiency of these reactors with respect to HSC proliferation. A flat-bed perfusion reactor (Fig. 6.2) was modified with the addition of grooves perpendicular to the direction of flow at the chamber bottom. Through a combination of experimental results and numerical simulations, the authors showed that a decrease in the spacing between grooves resulted in increase in reactor efficiency by as much as 25% without growth restriction. In an earlier report, after analysis of four geometries (slab, gondola, diamond, and radial shapes) for a parallel-plate bioreactor, the radial-flow-type bioreactor appeared to provide the most uniform environment for parenchymal cell growth and differentiation ex vivo due to the absence of walls that are parallel to the flow paths creating slow flowing regions. Such studies demonstrate the flexibility as well as the large margin for optimization on variations of perfusion chambers employed for HSC culture.

More recently, hematopoietic progenitors expanded in a perfusion bior-eactor were employed in a clinical trial for the treatment of breast

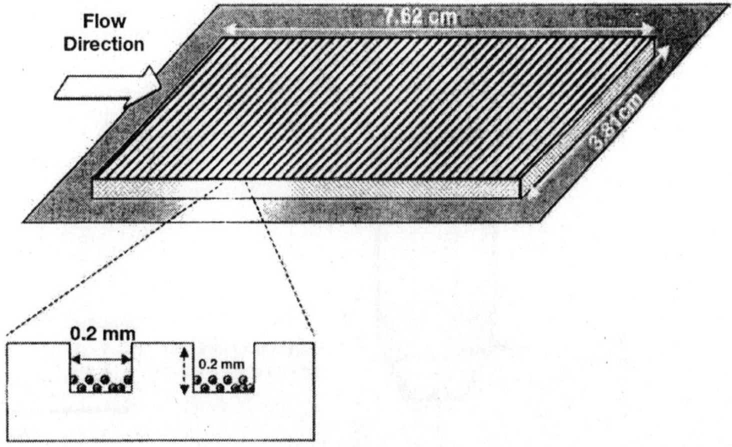

Fig. 6. 2 Flat-bed bioreactor with grooved bottom.

cancer patients. Cells were collected from patients, expanded and infused back to the patients, who in the mean time had undergone chemotherapy. The investigators observed platelet and neutrophile recovery after re-infusion suggesting the potential applicability of perfusion systems for stem cell culture in a clinically pertinent scale.

Although the results are encouraging, certain limitations of perfusion-based systems are also pointed out by these studies. Growth factor or oxygen transport constraints may curtail stem cell expansion. Surface area for cell growth poses a considerable limitation, a problem that can be exacerbated by the fact that high cell density favors cell differentiation instead of uncommitted propagation.

Other Bioreactor Systems

Except from stirred suspension vessels and perfusion chambers, a variety of other cell culture configurations has also been studied with the aim of large-scale stem cell expansion and differentiation. Sardonini and Wu examined static culture, suspension culture, microcarrier culture, airlift reactor, and hollow fiber culture. Using identical media, cytokines, and feed schedules, low density mononuclear cells in the suspension bioreactor expanded to a value of 1.6 compared to a normalized value of 1.0 for static cultures for the two runs investigated. Conversely, microcarrier and airlift bioreactor cultures averaged 0.75- and 0.70-expansion, respectively, compared to static cultures. However, in a later study, airlift packed bed bioreactor (Fig. 6.3) culture yielded promising

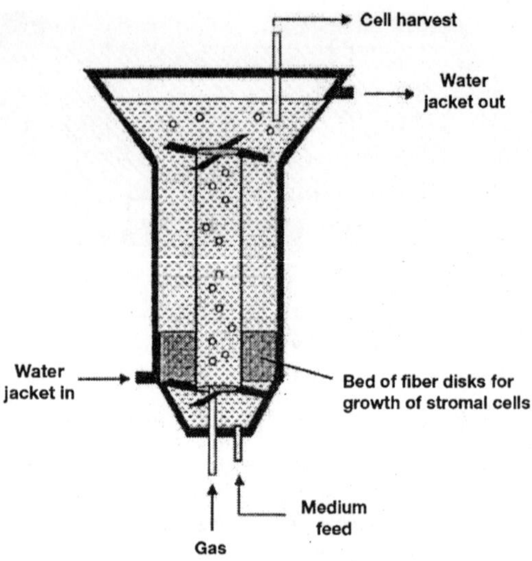

Fig. 6.3 Airlift packed bed bioreactor.

results in terms of ex vivo hematopoiesis. In this case, stro-mal cells were grown on a fiberglass matrix packed in the bottom of an airlift reactor, imitating the environment in the native bone marrow. Once the stromal cell layer was established, fresh bone marrow was inoculated and over an 11-week culture period approximately 3.6×10^8 cells were

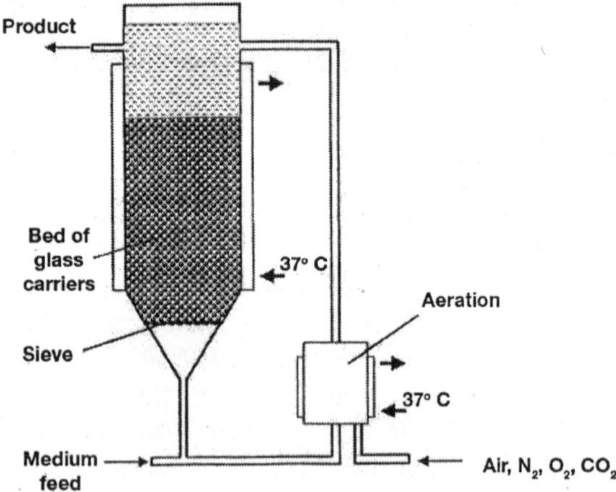

Fig. 6.4 Fixed bed boreactor.

obtained. Modifying a continuously perfused bioreactor, Meissner et al. developed a fixed bed reactor (Fig. 6.4) with stromal cells immobilized on porous glass carriers and cocultivated human hematopoietic progenitor cells for several weeks. Early progenitor cells expanded up to 4.2-fold whereas colony-forming unit-granulocyte monocyte and burst-forming unit-erythro-cyte up to 7- and 1.8-fold, respectively. Extensive production of lymphoid cells and all stages of committed cells as well as erythropoiesis were also achieved in bioreactors packed with porous beads in the presence of various cytokines.

These observations suggest that these systems have the potential to play a major role in realizing the promise of stem cell-based therapies. At the same time, they indicate that the design of bioreactor systems is crucial for efficient stem cell expansion.

Recent Issues in Adult Stem Cell Research

There is now a large body of evidence suggesting that tissue-specific stem cells are capable of adopting multiple fates. Claims of stem cells "transdifferentiation" or "plasticity" have been met with skepticism—if not criticism—as they challenge decades of experiments that have proved that animal cells once differentiated show little sign of haphazard fate-switching. Both proponents and opponents have come forward with various interpretations of the observed "transdetermination" phenomena. The former believe that the great plasticity of adult somatic stem cells may be expected because different niches share cytokine requirements, extracellular matrix components and transcriptional regulators. Considering this concept, many studies have illustrated the pleiotropic functions of growth factors and their receptors, adhesion molecules, and transcription factors.

However, two recent publications suggested that the apparent reprogramming of adult stem cells might instead be due to cell fusion. When bone marrow from GFP transgenic mice was cultured with ES cells, about 2-11 cells/10^6 cells fused with ES cells and these cells could subsequently adopt some of the phenotypes typical of ES cell differentiation. The cell fusion frequency was not greater in the hematopoietic fraction of bone marrow and this finding raises concerns since these cells were thought to be responsible for liver engraftment. In a parallel study, pyro-mycin-resistant GFP-labeled mouse neural stem cells were cocultured with hygromy-cin-resistant ES cells. Following selection in ES cell growth medium, the remaining cells expressed

GFP and were resistant to both pyromycin and hygromycin. Moreover, these cells had double the normal DNA content and showed many characteristics of ES cells. The data supported the occurrence of fusion at low frequency (10^{-5}) between CNS stem cells with ES cells and these fused cells exhibited multili-neage potential when injected into blastocysts, most prominently into liver.

These studies were a wake-up call for several groups and fueled efforts to establish rigorous standards for proving plasticity. By definition, stem cells are clonal precursors of more stem cells (symmctric division) and differentiated progeny (asymmetric division). Therefore, putative stem cells, which may be marked, should give rise to more stem cells starting from a single cell (clonality assay). Moreover, such cells should have the ability to repopulate and regenerate stem cells, progenitors and differentiated and functional progeny. Demonstrating the presence of tissue-specific proteins or morphology is not sufficient; the cells must contribute to the functions of the host tissue, e.g., detoxification functions for liver cells or electrophy-siological responses for heart or brain cells.

With these criteria in mind, many investigators are re-examining previously published results. For example, what was thought were muscle cells turning into blood-forming cells were actually hematopoietic cells already residing in muscle. To that end, little evidence was found of HSCs contributing appreciably, if any, to nonhematopoietic tissues including brain, kidney, gut, liver, and muscle. It is becoming clear that isolation and characterization of somatic stem cells are challenging tasks. Yet, dealing with these tasks may turn out to be easier through continuous advances in stem cell biology as well as remarkable progress in various cell-sorting techniques.

Potential Therapeutic Applications

The demand for stem cell-based therapies is great and grows continuously among different groups of patients. Degenerative diseases are developing into a frequent theme among the elderly as the average life expectancy has increased rapidly over the last decades. Cardiovascular and musculoskeletal diseases, diabetes, Alzeimer's, and Parkinson's pathologies are some of the most prominent conditions affecting millions of people even in western countries and future projections are bleak. The solution to these and other diseases may lay on stem cells in conjunction with novel bioengineering approaches such as tissue engineering. The moment when stem cells will be routinely used for treating human patients is probably not in the immediate future since numerous issues pertinent to stem cell biology are still under intense investigation.

Even so, results from a small number of trials have been reported while still others are underway and a large body of evidence exists from studies on various animal models. MSCs isolated from a bone marrow aspirate were expanded ex vivo and re-infused in breast cancer patients undergoing bone marrow transplantation. The results from a group of 28 patients indicated that the autologous MSCs effected early hematopoietic recovery based on platelet and neutrophil counts. In another study, MSC transplantation was employed to rectify skeletal abnormalities. Initially, transgenic mice with fragile bones due to osteogenesis imperfecta, a disease resulting from mutations in the type I collagen gene, underwent transplantation of wild-type MSCs. Bone collagen and minerals were restored partially in the recipient mice. Later, Horwitz et al. performed bone marrow transplantations from HLA-matched donors in three children with osteogenesis imperfecta. Engraftment was confirmed and was concomitant with new dense bone formation, increase in total body mineral content, and reduction of bone fracture frequency.

Except from MSCs, fetal neural cells which contain multipotent NSCs as well as more restricted neural precursors and terminally differentiated cells, have been employed for the treatment of Parkinson's and Huntington's syndromes. Fetal mesencephalic tissues were transplanted into the striatum of human Parkinson's patients. The grafted tissue survived for a long period in the human brain. Dopaminergic innervation to the striatum was restored and the patient had sustained improvement in motor function. In a similar fashion, the outcome of a clinical study with patients with Huntington's disease indicated increased scores on some measures of cognitive functioning following bilateral intrastriatal implantation of fetal striatal tissue. Although further investigation in this direction is necessary, the results suggest the possibility that purified populations of adult NSCs may be the key in reversing neurodegenerative conditions.

Stem cells can also be used as therapeutic agents in conjunction with gene delivery methods. The cells can be genetically modified to confer specific properties on their progeny. The ex vivo gene modification of stem cells intended for cell therapy reduces the possibility of unintentional carry-over of viral or other pathogenic vectors. Gene targeting methods, which are widely used to manipulate in a specific manner different chromosomal loci in ES cells, may be applied to adult stem cells as well. Furthermore, retroviruses, adenoviruses and adeno-associated viruses, which are typically used as gene transfer vehicles, exhibit tropism for various tissue-specific stem cells and provide additional means for ex vivo genetic manipulation of stem cells. MDSCs, or myoblasts, have been genetically engineered to deliver dystrophin to mdx mouse muscles, bone morphogenetic protein 2 (BMP2) for healing bone defects, growth hormone, ery-thropoietin, and factor IX. Human MSCs can be transduced with viruses and maintain transgene expression during expansion in culture and differentiation into cartilage, osteocytes, and adipocytes. Human skin stem cells have also been modified genetically to correct the expression of laminin in patients with severe junctional epidermolysis bullosa. Transplantation of neural stem cells retrovirally transduced to secrete nerve growth factor (NGF) was able to ameliorate the death of striatal projection neurons caused by transient focal ischemia in the adult rat.

A handful of gene therapy trial involved the use of hematopoietic cells for treating patients with conditions such as adenosine deaminase deficiency and Gaucher disease. Recently, Hacein-Bey-Abina et al. reported the use of genetically modified $CD34^+$ bone marrow cells for the correction of X-linked severe combined immunodeficiency due to mutation in the common γ chain *(γc)* gene. The study was conducted over a period of 2.5 years with the participation of five patients. The γc gene was transferred ex vivo via a retroviral vector to $CD34^+$ cells from these patients. Transduced T cells and natural killer cells appeared in the blood of four of the five patients within 4 months. Although the frequency of transduced B cells was low, serum immunoglobulin levels and antibody production after immunization were sufficient to avoid the need for intravenous immunoglobulin. Correction of the immunodeficiency eradicated established infections and allowed patients to have a normal life.

In a later correspondence, however, the authors commented that although three of the four patients continued to do well, 3.6 years after gene therapy, a serious adverse event occurred in the fourth patient. At a routine checkup 30 months after gene therapy, lymphocytosis consisting of a monoclonal population of $V\gamma9/V\delta1$, γ/δ T cells of mature phenotype was detected. Upon further examination, one proviral integration site within the *LMO-2* locus of chromosome 11 was found resulting in aberrant expression of *LMO-2,* which is linked to acute lympho-plastic leukemia. The increase in patient's lymphocyte count was accompanied by the development of hepatosplenomegaly. Although other possible explanations have not been excluded, these findings were interpreted as the consequence of the insertional mutagenesis event associated with the gene transfer using recombinant retrovirus. So far, the risk from such an event has been considered very low in humans.

As this and other studies on the development of stem cell therapies show, despite the promising results, important issues need to be addressed for successful application of these therapies in the clinical setting. The availability and expansion of the cells is obviously a major issue. Not only more refined methodologies for the isolation and purification of somatic stem cells are needed but the development of systems for in vitro maintenance, increase in cell number and directed differentiation of stem cells is essential. In the case of ex vivo gene delivery to stem cells, the gene transfer efficiency, the persistence of transgene expression over time and potential immunogenicity and/or oncogenicity of cells undergoing gene transfer are parameters which should be scrutinized extensively. Thus far, typical viral and nonviral vehicles used for gene therapy appear to be suitable for transduction/transfection of adult stem cells although further investigation is forthcoming. Transgene expression on the other hand is frequently dampened, for example, after retroviral transduction of stem cells. Silencing of a retroviral long-term repeat (LTR) promoter, which can result from methylation of the LTR sequence or positional inactivation can be addressed by substituting LTR elements with tissue-specific promoters/enhancers. In vivo, silencing may be linked to factors/signals originating in the extracellular milieu where the cells are homing. Not only cell homing but also the requirement for local or systemic expression of the gene product is an important determinant for choosing the route of cell delivery. For systemic expression, cells can be infused in the general circulation whereas examples of delivery for local protein expression include subcutaneous implantation of encapsulated or scaffold-attached cells. Regardless of the way the cells are delivered, thorough assessment of potential immunogenic or tumorgenic properties of stem cells intended for transplantation is required. As the aforementioned study by Hacein-Bey-Abina et al. exemplifies, the risk for insertional mutagenesis associated with retroviral gene delivery should not be overlooked as improbable. Such risk may be eliminated with the use of homologous recombination methodologies already applied for the genetic modification of ES cells.

Homologous recombination allows for better control over the site of integration but is hampered by low transfection efficiency, a problem amenable to optimization.

It is obvious that most studies involve animal models whereas data from stem cell therapy trials on human patients are sparser. Animal models certainly have limitations but they provide the best proving arena for potential cell therapies. Regardless of intrinsic differences between animal and human pathophysiology, data gathered from animal studies have tremendous predictive value for clinical translation of stem cell therapeutics in patients. In the clinical context, one envisions that the donor and the recipient of the adult stem cells are the same person and the importance of this is realized considering the minimization of serious immunorejec-tion problems. Furthermore, no study to this day has proved that adult stem cells are tumorogenic as ES cells which form teratomas upon transplantation. So far, however, the isolation and purification of adult stem cells is difficult and unlike ES cells, most tissue-specific stem cells do not appear to be pluripotent. Finally, the quality and characteristics of somatic stem cells may depend on a number of factors including the age of the donor.

7

Quality Control Procedures for Stem Cell Lines

The culture of human cells in vitro has provided important insights into cell biology, disease processes, and potential therapies. The advent of the culture of human embryonic stem cells has opened up an exciting new generation of possibilities including their potential for application to human regenerative medicine. However, in vitro cell culture brings a number of challenges; the cells and the cell culture environment are ideal for the growth of numerous microorganisms, and the cells themselves are prone to genetic changes. Furthermore, it is, unfortunately, common for cell cultures to be interchanged, cross-contaminated, or mislabeled during laboratory manipulations. This leads us to define three critical characteristics of cell cultures that are fundamental to the assurance of good-quality cell culture work. These are:

- Purity—The cells are free from microbiological contamination.
- Identity—The cells are what they are claimed to be.
- Stability—The genotype and phenotype remain stable during growth and passage in vitro.

Serial passage exposes cell cultures to the repeated risk of contamination by environmental microorganisms. Such contaminations are generally recognized by a significant change in the pH of the culture medium (as identified by a color shift in the medium) and the sudden appearance of turbidity or colonies of fungal organisms. In these situations the culture is generally not recoverable and should be discarded carefully to prevent contamination of other cultures. However, some microbial contaminants can establish subliminal persistent infections that are not obvious on visual inspection. These are commonly due to mycoplasma but may be caused by other organisms.

Mycoplasma contamination is known to cause a broad range of permanent and deleterious effects on cells including chromosomal abnormalities and cell transformation. Mycoplasma contamination is not obvious by visual inspection, can spread rapidly to other cultures handled by the same or other operators, and is difficult to eradicate reliably. Accordingly, it is important to perform routine screening for this organism in cell cultures.

The first human cell line HeLa, established in 1952, was widely distributed to laboratories also attempting to establish new cell lines. Within a few short years it became apparent that some of the "novel" cell lines being established were in fact cross-contaminated with HeLa cells. The problem was highlighted through the use of karyology and isoenzyme analysis but was only partially resolved and led to much scientific controversy. Since the identification of early cases of cross-contaminated cultures, cases have continued to be identified. Despite periodic reminders from concerned cell culturists the problem appears to continue, and recent publications seem to indicate that part of this problem may be due to cross-contamination at the source of the cultures in the laboratories of the originators of the cell lines. It is vital that this situation does not develop with stem cell lines, as this could cause confusion in laboratory experimentation and major problems in potential clinical application and hamper future development and acceptance of the technology.

One of the key issues in the development of hES cell lines has been the consistency and comparability among hES cells isolated at different centers under different conditions. Much work has been published on new culture and differentiation methods; however, each publication generally deals with a very limited number of cell lines. This leads hES cell researchers to ask whether the data produced can be applied more broadly to all hES cell lines. Approaches to technical standardization have been considered, and in addition one attempt to characterize hES cells on an international basis has been initiated. Generally, such attempts at standardization have been based on antibody markers developed for the study of early development in embryonal carcinoma models. Today a growing array of molecular and antibody markers for stem cells is developing that will be of use in the quality control of stem cell lines. In the following sections we explore the various techniques and methods that can be used to test stem cell lines to address issues of purity, identity, and stability and to qualify their use in stem cell research. Although much of this addresses hES cell lines, it is applicable equally to all cell lines whether derived from stem cells or not and whether embryonic, newborn, or adult.

The Cell Banking Principle

To limit the chances of contamination and genetic change it is wise to keep the number of passages of cells to a minimum. An important approach to achieve this, called the master/working bank principle, has been adopted in industry for many years. This prescribes the establishment of a *master cell bank* that will provide the reference point for future work with a cell line. This bank should be well characterized and subjected to quality control tests. Ampoules from the master bank are then used to produce larger *working cell banks* that can be made available for experimental purposes or distribution to other workers. The working cell bank should again be subjected to quality control, although this may be more limited and concerned mainly with identity and absence of contamination. If prepared correctly, this tiered master/ working bank system (Fig. 7.1) can provide reproducible and reliable supplies of identical cultures for research work over many decades.

The quality control tests that should performed as a matter of routine for all cell banks include viability (typically Trypan Blue dye

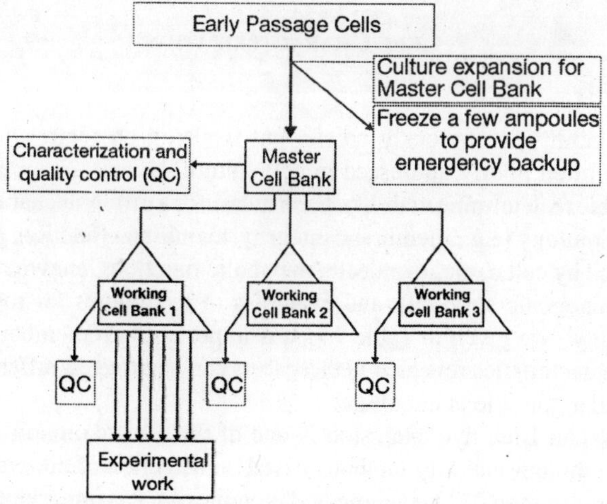

Fig. 7.1. A scheme for the establishment of master and working cell banks.

exclusion) and testing for absence of bacterial, fungal, and mycoplasmal contamination. These tests should be performed on cultures after a period of at least 5 days, and preferably two passages of antibiotic-free culture, to ensure that any contaminants that may be suppressed by antibiotics do not go undetected. Other tests for authenticity (e.g., karyology, DNA fingerprinting, isoenzyme analysis, surface markers) and assays for the presence of viruses may be performed, but the exact profile of tests will depend on the type of cells involved and the intended use of the cells. For a general reference on cell banking and quality control. The specific approaches and methods are discussed in the following sections.

Effective cryopreservation protocols are clearly essential for cell banking. Standard methodologies for other cell lines and for preservation of mouse embryonic stem cell lines have not been reported as being very successful with hES cells compared to vitrification methods. Most vitrification methods used for hES cells have been adapted from methods established for bovine oocytes and embryos, and the most commonly referenced modification used for hES cells. Successful methods reported generally utilize dimethyl sulfoxide and efhylene glycol as cryoprotectants, but details vary between publications. For a review of the methods currently used. Vitrification, while effective for preservation of hES cells, has a number of drawbacks including the need to carefully maintain storage temperatures close to liquid nitrogen temperatures to avoid devitrification, the costs of shipment under such storage conditions, the small volumes that can be frozen for each vial of cells required to be archived, and the rapid cooling rate required for successful vitrification.

Cell Characterization

Viability

Cell viability is obviously crucial but is also a characteristic that is all too often poorly addressed in cell culture. Numerous methods are available to determine viability. Each measures a different characteristic of cell biology (e.g., membrane integrity, membrane function, products released by cell damage or death, metabolic functions, enzyme activity, and clonogenic survival), and examples of techniques for measuring "viability" are given in Table 7.1. It is important to remember that the cell characteristics revealed in these tests can be affected differently by particular conditions in culture.

Trypan Blue dye exclusion is one of the most common methods used, although viability methods based on detection of apoptotic cells are also common. Whatever method is employed, it is important that it is

TABLE 7.1 VIABILITY TESTING FOR ANIMAL CELL CULTURES

Method	Principle and Comments
Dye exclusion (e.g., Trypan Blue, Naphthalene Black)	Dyes that penetrate cells are excluded by the action of the cell membrane in viable cells; thus cells containing no dye have functional membranes and are probably viable. **Advantages:** Rapid and usually easy to interpret **Disadvantage:** May overestimate viability since apoptotic cells continue to have active membranes and may appear viable.
Neutral red assay	Viable cells accumulate the red dye in lysosomes, and the dye incorporation is measured by spectrophotometric analysis. **Advantages:** Useful for certain toxicology assays **Disadvantages:** Time consuming and incubation conditions need to be optimized for each cell culture.
3-(4,5-dimethylthiazol-2-yl)-2,5-diphenyltetrazolium (MTT) assay	MTT reduction is measured by the formation of a colored product, and this is indicative of biochemical activity. **Advantages:** Many tests can be performed rapidly in 96-well array in automatic plate readers. **Disadvantages:** Some inhibited cells show a low MTT reduction value that is not necessarily related to cell viability.
Fluorescein diacetate assay	Fluorescein diacetate enters the cell and is degraded by intracellular esterases, releasing fluorescein that cannot escape from cells with intact membranes, and thus the cells fluoresce when observed under UV light. **Advantages:** Rapid setup **Disadvantages:** Requirement for UV microscope or flow cytometer

relevant to the cell culture application and that it provides reproducible results. For stem cell cultures it is clear that a viability measurement cannot predict the proportion of stem cells present in a culture after a culture treatment or cryopreservation, but frozen stocks of cells still can and should be checked promptly for viability by recovering a vial of cells into culture immediately after cryopreservation.

Karyology

Visualization of the cell's chromosomes (karyotypic analysis) provides a valuable perspective on the physical structure of the genome. It has been used as a valuable tool for monitoring the genetic stability of a cell culture and for recognizing the appearance of transformed cells, which are often aneuploid (having chromosome loss or duplication, or aberrant chromosomes with translocations, deletions, inversions, etc) and heteroploid (having a wide range of chromosome numbers per cell around or, more often, above the normal number).

T he method of visualization of chromosomes most commonly used today employs colchicine or a similar compound to block cell division at metaphase when the individual chromosomes are separate and condensed and thus most readily visualized. The cells are then harvested, subjected to swelling with hypotonic saline, KC1, or saline citrate, and fixed in acetic methanol before applying them dropwise onto microscope slides to create the characteristic chromosome "spreads." These are then stained with Giemsa to visualize the condensed chromosomes. The ability to identify chromosome pairs and to resolve the nature of fine alterations in chromosome structure was realized through the use of trypsinization before Giemsa staining, which reveals banding patterns characteristic of each chromosome. Other techniques for studying karyology have been developed, such as Q and R banding and chromosome painting, but the Giemsa banding method described is the most widely applicable and generally useful method that has been used to characterize various stem cell culture systems. A typical method and review of other methods.

More recently, studies of hES cell cultures have revealed that they are prone to karyological changes, and a major challenge has emerged in maintaining the cells in the undifferentiated state while preserving a diploid karyotype. No culture system has been able to prevent completely the tendency of hESC lines to accumulate karyotypic abnormalities. It is felt that, owing to the nonideal culture systems for hESC lines, selection pressure is present in favor of chromosome duplications that confer an adaptive benefit. For hES cells there appear to be common patterns of chromosome alteration representing "adaptation" of these cells to in vitro culture conditions, notably changes involving chromosomes 12 and 17. An example of a karyotype of a normal hES cell line is shown in Fig. 7.2.

It is important to verify that frozen stocks of cells retain the diploid karyotype and to check cells in use periodically to ensure that cells used to generate data for publication are diploid, unless deliberately studying transformed cultures. Many cytogenetics labs and some contract testing companies will provide a testing service for karyotype analysis. Routine diagnostic cytogenetic tests performed for clinical purposes often give analysis of 10-20 metaphase spreads. While such testing will identify the appearance of a transformed clone that is dominating the culture, it may not detect a low level of transformants occurring at the early stages of culture instability. It is often desirable to count at least 50 metaphase spreads in order to detect (or rule out) lower rates of mosaicism.

There are ongoing efforts to develop a chip-based or molecular assay for the kary-otypic stability of hES cells in culture. One of these

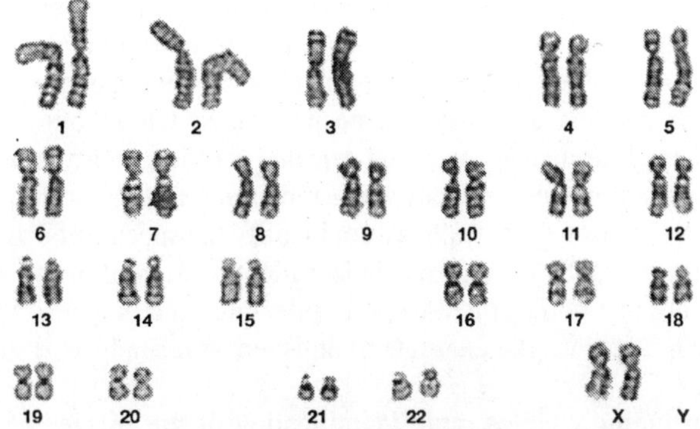

Fig. 7.2. Diploid human ES cell karyotype

methods is based on single nucleotide polymorphism (SNP) genotyping. SNP arrays are very useful in mapping markers of genetic diseases and for detecting loss of heterozygosity (LOH) in cancer. Technological advances now enable the use of oligonucleotide SNP arrays to measure chromosomal copy number at high resolution. This expands the utility of SNPs to detect nonreciprocal translocations, aneuploidy, and partial amplifications or deletions of chromosomes, and even amplifications or deletions of small chromosomal regions. The SNP array method has some advantages over conventional methods, mostly based on the resolution and size of genomic changes that can be detected. Based on a minimal detectable signal from 10 SNP sequences, currently available arrays of 550,000 SNPs have an effective resolution of about 28 kb, an array of 660,000 SNPs has an effective resolution of 25 kb, and, of course, increased density arrays that are sure to enter the field soon will improve this even further. One limitation that must be kept in mind, however, is the deficiency of molecular methods in analyzing heterogeneous or mosaic cell populations.

Identity Testing

Confirmation of Species of Origin

Isoenzyme analysis is based on measuring the charge-to-mass ratio of different isoenzyme activities using an agarose-based gel to separate the various polymorphic enzymes that can be identified for even just one enzyme reactivity. The cells are lysed, and the released enzymes are stabilized in a buffer. Samples of this preparation are then subjected to agarose electrophoresis before the gel is treated with a specific enzyme substrate and the reaction is visualized by the formation of a purple formazan product. This method has been made more reliable with the advent of a commercial kit (AuthentiKit, Innovative Chemistry), and testing for specific enzymes can be performed within one working day. The enzymes usually used are selected for their ability to identify polymorphism between species while remaining monomorphic within species. A single enzyme test may not identify the species of origin, but certain enzyme substrates provide clear identification of the species of origin using just two or three enzyme substrates and may also identify embryonic isoforms. An example of an isoenzyme analysis is given in Fig. 7.3.

Such typing enables rapid identification of the species of origin within one working day, and depending on the enzyme substrates used it may also allow the identification of the strain of origin for mouse cell lines. Such levels of differentiation will be valuable in a laboratory using cells from diverse species but may not be so useful in a laboratory that only cultures human cell lines.

Numerous molecular methods are now available for confirming the species of origin based on the amplification by the polymerase chain reaction (PCR) of conserved sequences and sequencing of specific genes such as cyto-chrome oxidase. The latter method provides a sequence specific to each species that is supported by a growing database of sequence data maintained by the US National Center for Biotechnology Information and may well become a reference method for species identification.

Species identification is a useful part of cell authentication, and which method is used will be a decision based on the types of cell lines used in the laboratory, staff time available to carry out in-house testing, and access to appropriate facilities and equipment.

DNA Profiling for Cell-Specific Identification

Variable number tandem repeats (VNTRs) and short tandem repeats (STRs) are interesting sequences in the human genome that are comprised of repeated core units of sequences, some of which, when excised from the human genome with certain restriction enzymes, show polymorphism between individuals in the number of repeat units at a particular genomic locus. It was Alec Jeffreys who discovered that this variation might be used to identify and discriminate between human individuals by means of certain genomic probes and Southern blotting

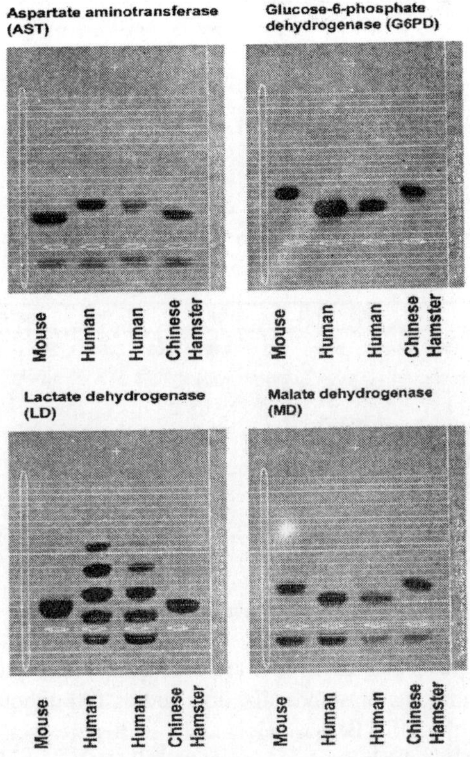

Fig. 7.3. Isoenzyme profiles for cells from mouse, human, and Chinese hamster cell lines.

following electrophoresis. Other workers identified similar probes based on other VNTR sequences.

Application of methods based on the hybridization of various probes to Southern blots of cell line DNA developed rapidly in the 1990s, and quickly these powerful methods, including PCR-based DNA profiling, revealed numerous cases of cross-contamination.

STR loci consist of short, repetitive sequences, 3-7 base pairs in length. These repeats are well distributed throughout the human genome and are a rich source of highly polymorphic markers, which are routinely detected with PCR. Alleles of STR loci are differentiated by the number of copies of the repeat sequence contained within the amplified region and are distinguished from one another with radioactive, silver stain, or fluorescence detection after electrophoretic separation. Commercial kits that contain labeled primers to detect the

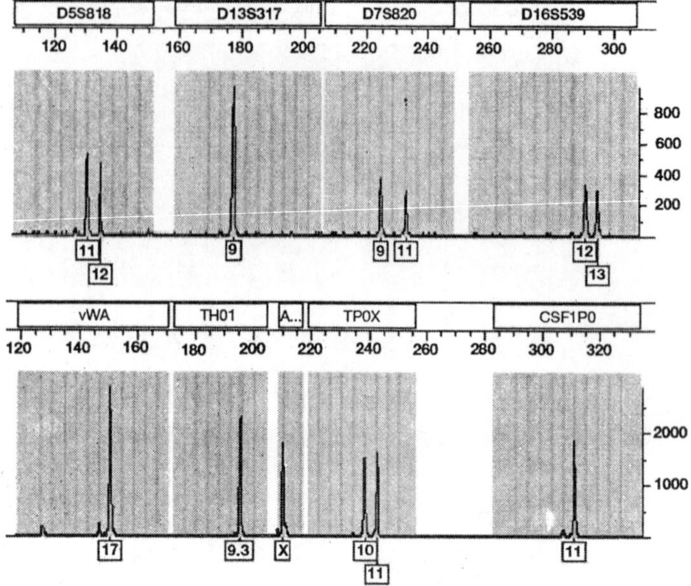

Fig. 7.4. STR electropherograms.

number of repeats at 8 or 16 loci are available. These loci satisfy the needs of several major standardization bodies throughout the world; for example, the US FBI has selected 13 STR core loci from the set of 16 to search or include samples in CODIS (COmbined DNA Index System), the US national database of profiles of convicted offenders. The matching probability of the 8 loci system ranges from 1 in 1.15×10^8 for Caucasian Americans to 1 in 2.77×10^8 for African-Americans, while the 16 loci system ranges from 1 in 1.83×10^{17} for Caucasian Americans to 1 in 1.41×10^{18} for African-Americans. A sample STR electropherogram is shown in Fig. 7.4.

Antibody Markers

An important characteristic of any cell is its profile of antigen expression. A panel of antibodies has been commonly used to characterize hES cell antigens and show typical patterns of reactivity in such cultures (Table 7.2). Characterization of hES cells by immunophenotyping is best performed qualitatively by using immunohistochemistry together with quantitative analysis using flow cytometry. Several current markers are largely based on a single precursor (lactosylceramide) that undergoes biochemical modification including glycosylation to create the different epitopes representing the stage-specific early antigens (SSEAs). Some of these markers have proven useful for tracking the differentiation of hES cells and are key identifiers for hES cells, although they are not unique to this cell type. New markers are needed that have a direct functional relationship to stem cell biology.

It is important to take some care when sourcing antibodies for the characterization of stem cell lines. Each antibody used must be of the correct type, specificity, and titer, as described by the supplier. Fundamental approaches to this have been described elsewhere, but at the very least new stocks of antibody should be compared in parallel with existing acceptable stocks and antibody isotype controls should be included in experimental work. It is important to differentiate between antibodies that might be involved in the detailed characterization of stem cell lines. Table 7.3 identifies a number of antibody markers that might be used. In characterization of master stocks and for publications, comprehensive antigenic characterization may be required. In the case of the UK Stem Cell Bank, the panel of markers such as those highlighted in Table 7.3 are applied to both master and distribution stocks to ensure that the material released to researchers is of an acceptable quality. Some quality control on cultures used for experimental work is clearly desirable, but it is not practical to carry out detailed profiling in this situation. Observation of hES cell morphology may give an indication of the state of differentiation in routine observation of experimental and stock cultures, but it could be valuable to have flow cytometric data on a marker of differentiation, such as SSEA-1, at critical points in the use of cultures.

Gene Expression

The study of gene expression profiles is extremely valuable but is largely just developing at a research level. The application of such tests in routine work has yet to be validated, but there are a number of genes associated with sternness that should be checked (e.g., Nanog, Oct-4) as well as testing examples of genes that are associated with differentiation. Care should be taken to avoid primers that also detect pseudogenes and could cause confusion. At this point in time it is probably not wise to set a specific panel of genes for quality control until better culture methods and knowledge of the nature of stem cells are obtained. However, it will clearly be useful to gather information on gene expression profiles on candidate stem cell genes and markers of different differentiation lineages. Examples of genes being analyzed

Table 7.3 Typical Antigenic Marker Expression of Hes Cells

Antigen	Typical Reaction on Undifferentiated hES Cells	Typical Reaction on differentiated hES Cells
SSEA-1	−	+
SSEA-3	+	+*
SSEA-4	+	+*
Oct-4	+	+*
Alkaline phosphatase	+	+*
TRA-181	+	+*
TRA-160	+	+*

in parallel in a current international study of over 60 hES cell lines are given at www.stemcellforum.org.uk/.

Recent research comparing the transcriptomes of multiple hES cell lines has identified a set of approximately 100 genes that are highly expressed in undifferentiated hES cells. In a study characterizing 17 different hES cell lines by whole genome gene expression array analysis, almost all of the cell lines show a similar expression pattern for these 100 genes. While global gene expression can be indicative of the state of differentiation of a particular hES cell line, it is most useful when compared side-by-side with a reference standard for the particular differentiation state under investigation. For example, human embryonal carcinoma cell lines and stable, karyotypically abnormal hES cell lines have been proposed as such standards. Microarray expression data are best verified by quantitative real-time reverse transcriptase PCR (RT-PCR), using a marker set for genes commonly associated with the hES cell differentiation state of interest. For the undifferentiated state, a few accepted markers are listed in Table 7.3.

Pluripotency

This is clearly a key measure of stem cell line performance in which the expected outcomes may vary depending on the cell type (hES, mesenchymal stem cells, etc.). There are a number of ways of measuring pluripotency including the following:

- Germ line competence (only acceptable for nonhuman stem cells)
- Teratoma formation in immunocompromized mice
- Generation of embryoid bodies with the three germ layers represented
- Differentiation in vitro into cell types representing the three germ layers.

In hES cell research, teratoma formation is generally accepted but is not yet well standardized—molecular assays of gene expression may help in the future. This is clearly a challenging area that requires considerable research effort before reliable QC methods can be selected with confidence. (*See* Chapter 6 for protocols and a discussion of teratoma formation.)

Briefly, undifferentiated hES cells are injected into immunodeficient mice. SCID/Beige is a commonly used strain. The cells are placed either intramuscularly into the hindlimb, in the testis, or under the kidney capsule. There have not been careful comparisons of injection locations, but all three mentioned above appear permissive for pluripotent hESC differentiation. Tumors form in all mice and are excised for analysis after approximately 8-12 weeks, based on tumor size. The first analysis performed is histological, preferably performed by an experienced pathologist. In this analysis, the pathologist attempts to identify tissues representative of the three germ layers (ectoderm, endoderm, and meso-derm). This may be followed by immunohistochemical and gene expression analyses to demonstrate that the teratomas contain specific terminally differentiated lineages. Typical teratomas are generally well demarcated from the host tissue and exhibit organized clusters of cells, which may include cartilage, mineralized bone, villi, smooth muscle, nerve bundles, neural rosettes, liverlike structures, ducts, cystic epithelium-lined spaces, and various types of epithelia.

Sterility

Bacterial and fungal contamination generally prevents work with affected cultures as they become turbid with organisms that completely overwhelm and kill the cells. Contamination can arise from a variety of sources in the laboratory environment (e.g., water baths, fridges, sinks, cardboard boxes), and avoidance of contamination is most effectively achieved with good aseptic technique, correct use of class II safety cabinets (see Appendix 1), and maintenance of a clean and tidy cell culture laboratory. The use of antibiotics may be helpful to avoid loss of cells in circumstances where the risk of contamination is high, for example, in primary mouse embryonic feeder cultures or in routine experimental work where environmental contamination is very high. However, routine use of antibiotics for cultures that should be "clean" is not recommended and certainly not for the preparation of cell banks. Antibiotics can affect the function of cells, and routine use of specific antibiotics can encourage the development of resistance in microorganisms, leaving no fallback treatment for protection of critical cultures. They may also suppress but not eliminate the growth of mycoplasma, increasing the risk of false-negative results in mycoplasma testing.

Cultures can be tested by inoculation of the supernatant medium from a culture into bacteriological broth followed by incubation at both standard cell culture temperature (typically 35-37°C) and room temperature to reveal growth of contaminants with different optimal growth temperatures (see scheme in Fig. 7.5). Reliance on recognition of contamination by appearance of turbidity in inoculated broths can lead to difficulties because nonmotile organisms may not produce turbidity, while cell debris may cause turbidity. Accordingly, it is usual to subculture all broths onto solid nutrient agar media at the termination of their incubation period to detect any culturable organisms.

Detailed reference methods for this approach are published in national pharmacopoeia.

In addition to these standard culture methods, there are a number of kits available from tissue culture companies that may have useful applications in stem cell work. Such methods may also have valuable application alongside traditional culture methods because some are more rapid and may detect more fastidious organisms that can arise

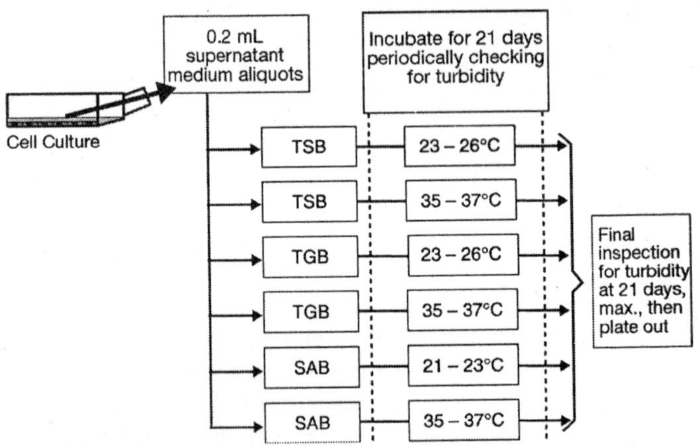

Fig. 7.5. Example of a typical sterility test. TSB, trypticase soy broth used to isolate aerobic and facultative aerobic organisms. TGB, thioglycollate broth used to isolate anaerobic and microaerophilic organisms. SAB, Sabourard's broth used for isolation of fungi. Alternative media could include Todd-Hewitt broth (instead of TSB), brain-heart infusion broth (instead of TGB), and YM broth (instead of SAB). Additional media may be added that contain blood or serum (e.g., nutrient agar incorporating 5% deflbrinated rabbit blood).

in cell cultures but would not be detected by the standard sterility tests described above. However, the range of organisms that could potentially be isolated in broth and agar cultures can be significantly expanded by supplementing these tests with additional growth media incubated in a CO_2-containing atmosphere. By far the most common fastidious contaminants are mycoplasma species, and specific tests for these organisms are discussed in the next section.

MYCOPLASMA TESTING

Mycoplasmas are organisms of the order *Mollicutes,* which are much smaller than typical bacteria and, while similar to bacteria, have a number of distinct characteristics that give them special potential to cause problems in cell culture work. They have a degree of resistance to the antibiotics normally used in cell culture and can pass through standard bacteriological filters. Furthermore, they do not necessarily affect the growth rate of cells and do not usually produce visual turbidity in the supernatant medium of a contaminated culture, and thus may go unnoticed. Persistent contamination with mycoplasma can cause a diverse range of permanent deleterious effects on cell lines, and it is vital to carry out routine screening of new cell cultures coming into the laboratory to avoid potential spread to other cultures.

There are a number of techniques for mycoplasma detection that have been used widely, and examples are given in Table 7.3. The reference methods used in industry are Hoechst 33258 staining of Vero cells inoculated with culture supernate and culture using selective broth and agar media. For routine screening, direct PCR or Hoechst 33258 staining (*see* Table 7.4) are useful, but these methods are generally not as sensitive in routine use as culture, which, in the absence of a more sensitive method, should be used for any important frozen stocks of cells that will be needed for future use. A scheme for a typical culture method and Hoechst 33258 staining is given in Table 7.4, which shows the added benefit of dual testing to give early screening results that will be confirmed some time later by the more sensitive culture method. It should be noted, however, that a considerable degree of skill is required to culture mycoplasma, and the need to include a positive control may not be acceptable in some laboratories without proper quarantine facilities. There are also a variety of proprietary methods available on the market, but, as indicated for sterility test kits, it is important to test these against a standard proven methodology before putting full confidence in them. This is important because the long-term consequences of missing

TABLE 7.4 COMPARISON OF DIFFERENT METHODS FOR DETECTION OF MYCOPLASMA

Technique	Advantages	Disadvantages
Broth and agar subculture	Highly sensitive	Bacteria may grow on selective media
	Well-established method	Will not detect nonculturable strains
	Standard methods available in national pharmacopoeia	Long incubation periods (approx. 50 days total)
Vero cell culture inoculation and DNA stain	Results in 3 days Standard method available in national pharmacopoeia.	Vero test cells must be maintained and prepared High-power (x 100 objective) UV-fluorescence microscopy required. Nuclear fragments from cells and small bacteria may give false positives with inexperienced workers.
PCR	Results within 1 day Large numbers of samples readily screened	Sensitivity needs to be demonstrated and monitored carefully, Nested PCR may give rise to false positives.
6-Methylpurine deoxyriboside (6-MPDR) Added to sample and indicator culture (e.g., 3T3, Vero). Mycoplasma contamination detected due to mycoplasmal adenosine phosphorylase. Converts 6-MPDR to toxic metabolites that kill indicator cells.	Simple end point (cell death)	Indicator cells must be maintained and prepared. Five days incubation required. False negatives have been observed when compared with other methods [e.g., Uphoff et al., 1992].
Mycoplasma RNA hybridization.	Sensitivity reported to be high but may vary.	Radioactive versions require scintillation counting equipment Difficult to discriminate between negative and low positive results.

positive cultures can be catastrophic in terms of wasted technical time and effort and invalidation of scientific data.

OTHER MICROBIAL CONTAMINANTS AND POTENTIAL BIOHAZARDS

A broad range of microorganisms could potentially contaminate human stem cell lines, and it would be impractical and too expensive to screen

TABLE 7.5 OUTLINES OF TYPICAL PROTOCOLS FOR REFERENCE METHODS FOR DETECTION OF MYCOPLASMA

Time (days)	Broth and Agar Culture	Hoechst stain
0	1. Inoculate 200-μL samples of supernatant medium into broth and onto an agar plate (contains thallous acetate and pig serum) and incubate in a anaerobic environment.	1. Replace medium on Vero cell monolayer (on glass coverslip) with supernatant medium to be tested and incubate in 5% CO2:95% air.
Day 3		2. Remove medium and fix monolayer with fixative (1:3 acetic acidmethanol) for 3 min. Replace with fresh fixative for a further 3 min. 3. Drain fixative, air-dry slides and immerse in 2 mL of stain (0.1 μg/mL bisbenzimide Hoechst 33258 in Hanks' BSS without Phenol Red or PBSA) and incubate for 5 min at room temperature in the dark. 4. Remove stain, add nonfluorescent mountant, and apply a coverslip. 5. Scan the stained area (100 ×epifluorescence) for fluorescent cell nuclei (acts as a control for stain) and for small fluorescing particles over the cytoplasm.
Days 3-5	2. Inoculate 0.2 mL of sample from broth to a selective agar plate and observe original plate.	
Day 14	3. Inoculate selective agar plate from broth and observe plates inoculated at day 0 and days 3-5.	
Day 21	4. Check for pH change in broth and subculture if altered. Observe all plates and return to incubation conditions with final observation at 28 days for each plate.	
Day 42	5. Observe any remaining plates for colonies.	

all cell lines for all of these organisms. While the future use of stem cell lines for therapy would require intensive investigation for microbial contamination, the use of these cells for research purposes should be based on sensible precautions that apply to any unscreened human cells. These precautions include a risk assessment based on any information available on the cells, use of aseptic technique, and good cell culture practice. Generally speaking, the risk associated with such cultures will be very low; however, there is no room for complacency because these cells could potentially carry a range of viruses due to their origin in the human reproductive tract (e.g., herpes virus, HIV, hepatitis B), and a careful combination of risk assessment (to avoid use of cells with a significantly raised risk of contamination with serious human pathogens), containment (e.g., use of sealed culture vessels, use of a Class II safety cabinet, treatment of cell culture waste as if infectious), and quarantine of cell cultures newly arrived in the laboratory until evaluations and basic quality control have been carried out should lower the risk.

Quality Control of Culture Conditions, Reagents, and Media

In cultures of stem cells there is great potential for variability and instability. It is helpful in dealing with these issues to try to control the variation in the nutritional and environmental influences to which the cells are subjected. Calibration and monitoring of temperature and CO_2 levels are clearly important, especially in multi-user labs, where incubators may rarely reach standard culture conditions during the working day. Use of consistent sources and consistent composition of key media and supplements is also important. In addition, any critical reagents most likely to suffer from batch-to-batch variation, such as bovine serum and serum replacement formulations, should be tested before routine use, and a batch reserved, to ensure they provide acceptable growth of cultures.

It is also important to remember that feeder cells are also a potential cause of contamination and this is a particularly high risk with primary cell feeder cultures. To avoid this significant risk of contamination it is wise to establish large stocks of cryopreserved feeder cell preparations that can be quality controlled (i.e., viability, sterility, and mycoplasma as a minimum) before use.

8

Biochemistry of Cells in Culture

Cells *in vivo* grow in a highly controlled and complex environment. On removal from their location *in situ* they are dispersed from a bistological configuration, where cell contact plays an important role, into a simplified growth medium. This lacks hormonal and neuronal regulatory mechanisms and the thousands of intermediate metabolites present in body fluids. In addition the organism has highly evolved oxygenation and detoxification systems ensuring that, except in conditions of stress, an extremely uniform environment is maintained. Thus a cell in culture is in a very alien situation, the stresses of which can only be minimized by attention to environmental optimization. An important factor is the nutritional environment, which has to replace the complex body fluids and will affect all the physiological and metabolic events in the cell. Cell media were first developed with components identified in body fluids (amino acids, vitamins, etc.), plus serum, to provide the multitude of unknown metabolites. Over the years the need for serum has been reduced, or replaced, as a greater understanding of the cells' nutritional requirements has been obtained. Thus, a whole range of growth factors, hormones and inorganic ions can be added in place of serum to give a more standardized and optimized environment. This in turn not only means greater cell and product productivity, but allows more informative studies to be carried out on understanding cellular regulation of growth and metabolism. This allows greater maximization of performance in culture, whether it is to enhance secretion of monoclonal antibodies or to respond specifically to a drug in a toxicity assay. Cells in culture no longer have to be de-differentiated. Thus, the ubiquitous cell can, with the correct environment, demonstrate many differentiated characteristics.

An understanding of cell biochemistry is essential to potentiate the use of cells for all purposes. Many nutrients have been identified as essential, as have many that play a key role in cell behaviour (e.g. glutamine, glucose, oxygen). Inhibitory factors such as ammonia and lactate have also been identified. However, it must be stressed that this is only the 'tip of the iceberg' when it comes to controlling the biochemistry of the cell. The macrometabolites may have been identified, but cellular control and regulation is based on a cascade of signals from micrometab-olites acting on the cell surface and in the cytoplasm. This is why studies at the molecular biology level of the role of growth factors in this cascade are essential for a full understanding of cell biochemistry.

The nutrient balance is not the only factor affecting cell biochemistry, as there are many physiochemical interactions in a culture. The need to scale up cell cultures increases the importance of understanding the effect of such factors as agitation and mixing, and of how oxygen should be supplied in the least damaging way.

Keeping the cells healthy and viable has always been a dominant priority. The recognition that cell death may be programmed or behavioural (apoptosis) rather than just the result of hostile environmental factors causing necrosis and lysis has opened up another means of controlling cell viability and longevity in culture. An understanding of the gene products that control these processes is gradually unfolding, leading to great expectations of improvements in our capabilities of controlling cellular functions.

It is easy to accumulate masses of metabolic data on cell performance in culture, especially at the level of oxygen, glucose and glutamine utilization. The difficult part is analysing and interpreting the data, as shown by the numerous publications giving voluminous data but totally lacking any form of subjective analysis. It is our aim to cover this complex and expanding area by looking at, for example, computer modelling of cell kinetics, in order to stimulate new approaches.

It is expected that a combination of environmental control, medium design, the control of growth factor-regulated gene expression and genetic modification of the cell will enable huge advances to be made in animal cell productivity.

Quantitative Analysis of Cell Growth, Metabolism and Product Formation

With the growth of the biotechnology industry there has been an ever-increasing volume of data obtained from a variety of cell types cultured under different conditions and in different culture systems. Unfortunately, the lack of standardization for data presentation makes it difficult to compare results. The frequently presented profiles of product or metabolite concentration versus time have limited value because the concentration depends directly upon the viable cell concentration. It is more useful to present results as production or consumption rates per cell (or unit mass). For example, the characteristic parameter for glucose consumption is the specific (per cell) glucose consumption rate (mmol glucose viable cell^{-1} unit time^{-1}) rather than the glucose concentration profile. The concentration of a product or nutrient is still important because it may directly affect the production or consumption rate of itself or other metabolites. For example, the glucose consumption rate increases with glucose concentration (up to 1-5 mM). Glucose consumption is also a function of culture P_{O2} and pH. Thus it may be necessary to control parameters such as pH, P_{O2} and glucose concentration in order to examine the effects of these or other parameters (e.g. hormone levels) on cell metabolism. At a minimum it is important to report the levels of these and other parameters, such as ammonia and glutamine concentration, when presenting experimental results.

Typical commercial cell culture systems include batch or fed-batch suspension reactors and perfused immobilized-cell reactors. However, the transient nature of batch culture causes difficulties in studying the effects of external stimuli on growth, metabolism and product formation. Due to metabolite concentration gradients, and the difficulty of obtaining representative cell samples, immobilized-cell reactors are also poorly suited for the analysis of cell growth and metabolism. As a result it is desirable to use well-defined model systems. Continuous-flow suspension reactors allow metabolic parameters to be measured at steady state, after cells have adapted to new (and possibly inhibitory) conditions. Perfusion reactors (with cells immobilized on suspended or stationary supports) extend these benefits to anchorage-dependent cells, and provide model systems for cell responses *in vivo*. However, while it is instructive to study the behaviour of cells under well-defined conditions, the results obtained must be verified in the culture system selected for commercial production.

In this section methods are introduced for processing and presenting data for cell growth, metabolite consumption and product formation in a manner that facilitates

comparison between diverse reactor systems and cell types. Data analysed in this manner also facilitate application of results obtained in one reactor system to another. Equations are presented for a variety of culture configurations, such as batch, fed-batch, continuous and perfusion of immobilized cells. Specific examples are included to illustrate the methods of analysis.

Errors in Calculations

Experimental errors tend to be quite large in biological systems, e.g. ±30% in protein concentration measurements. Cell number measurements are generally no better than ±5%. At lower viabilities (<70% viable), accurate determination of viability and cell number is difficult, and the error in each determination may be greater than ±10%. Errors in cell and metabolite concentration measurements lead to uncertainties in calculated parameters, such as specific growth, production and consumption rates, therefore a complex profile for these calculated parameters should not be assumed when a straight-line or simple function will suffice.

Errors for batch culture tend to be larger than those for continuous culture due to rapidly changing conditions and limited samples. When dealing with batch cultures, as many samples as possible should be taken; however, care should be taken that the culture volume is not depleted so much that the culture conditions are affected (e.g. K_La as discussed below). Continuous culture allows for more frequent samples. However, the total sample rate should not exceed the dilution rate so as to maintain a constant reactor volume.

Cell Growth And Death Rates

The central parameter in cell culture is the viable cell concentration. The specific growth rate (μ), the specific death rate (k_d) and the fraction of viable cells (f_v) are used to characterize the proliferative state and health of a culture. These parameters are calculated from the viable and non-viable cell concentrations using equations appropriate for the type of culture vessel employed. The viable cell concentration is also used to calculate the metabolic quotients (e.g. specific glucose consumption rate), as discussed in a later section. In the absence of cell concentration data, analysis is limited to ratios of metabolic rates.

Free cells in suspension

General schematic of a well-mixed suspension culture vessel. For semi-batch culture the outlet flow rate would be zero, while for batch culture both flow rates would be zero. Assuming constant density, a total mass balance around the vessel is:

$$\frac{\mathrm{d}V}{\mathrm{d}t} = F_\mathrm{i} - F_\mathrm{o}$$

where V is the volume of medium in the vessel, F_i is the volumetric flow rate of fresh medium and F_o is the volumetric flow rate of the spent medium (including suspended cells). The balance on total cells is given by: rate of accumulation of total cells = rate at which cells are added in the feed − rate at which cells are removed in the outlet + rate of cell growth − rate of cell lysis.

The balance on total cells for a sterile feed is given by:

$$\frac{\mathrm{d}(nV)}{\mathrm{d}t} = 0 - nF_\mathrm{o} = \mu_\mathrm{app} nV$$

where n is the total cell concentration and μ_app is the apparent specific growth rate, which includes cell growth and lysis.

Noting that only viable cells can divide and assuming that viable cells are not directly lysed, Equation is obtained:

$$\mu_\mathrm{app} n = \mu n_v - k_\mathrm{l} n_\mathrm{d}$$

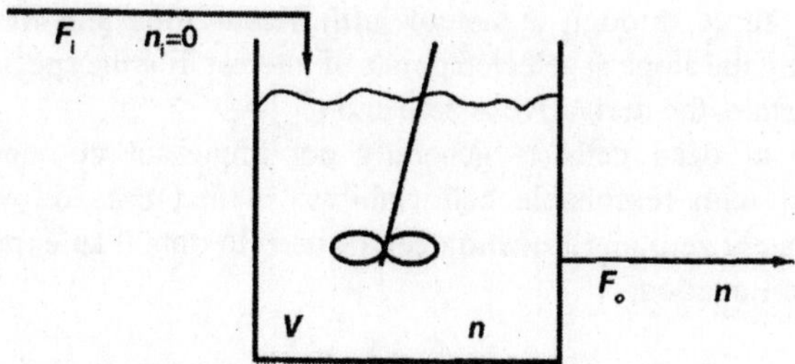

Fig. 8.1 Schematic diagram of a continuous-flow stirred suspension vessel. For semi-batch culture $F_o = 0$ and for batch culture $F_o = F_i = 0$.

where μ is the true specific growth rate, n_v is the viable cell concentration, n_d is the dead cell concentration and k_l is the dead cell specific lysis rate. Note that cell death converts a viable cell into a dead cell, but does not directly change the total cell concentration. The viable cell balance is given by:

$$\frac{d(n_V V)}{dt} = 0 - n_v F_o = \mu_{app} n_v V$$

μ_{app} is the viable cell-derived apparent specific growth rate given by: where k_d is the specific death rate.

$$\mu_{app} n_v = \mu n_v - k_d n_v$$

Continuous culture

For a constant-volume vessel $F_i = F_o = F$. Significant acid and base additions for pH control should be included in the F_i term. Reactor samples are included in the F_o term and do not affect the analysis as long as V is approximately constant. Equation becomes:

$$\frac{dn}{dt} = -nD + \mu_{app} n$$

where the dilution rate $D = F/V$. Equation can be rearranged to obtain a general expression for μ_{app} as in Equation:

$$\mu_{app} = \frac{d\ln(n)}{dt} + D$$

Because cell count data are often noisy, it is best to obtain the derivative by fitting ln(n) versus time with a low-order polynomial, or by drawing a smooth curve through a plot of ln(n) versus time and manually determining the slope at selected points of interest. For the special case of steady state, the derivative is zero and $\mu_{app} = D$.

Lysis of dead cells is generally not important at conditions of interest with reasonable cell viability. In that case &, will be approximately zero and Equation can be used to obtain an expression for μ as in Equation:

$$\mu = \mu_{app}\left(\frac{n}{n_v}\right) = \frac{\mu_{app}}{f_v}$$

Note that u, diverges from μ_{app} (and hence D at steady state) as the viability decreases. This is because new cell growth must offset cell death, as well as cell removal in the outlet stream. The cell death rate can be obtained from Equations:

$$k_d = \mu - D - \frac{d\ln(n_v)}{dt}$$

The death rate is generally low at high dilution rates and becomes more significant at low dilution rates.

Example 1

Table 8.1 contains cell concentration and viability data for a continuous culture experiment in which a pH step change was implemented. The initial steady state was established at pH 7.6. At time zero the pH was decreased to 7.1. A plot of the natural logarithm of total cell concentration versus time is shown in Fig. 8.2a. A slope of zero is assumed for the initial (time < 0 days) and final (time > 7 days) steady states. A constant slope is taken for the exponential growth portion (1 day $<$ time $<$ 5.5 days). Two other slopes are also illustrated in the Fig. The calculated slopes are substituted into Equation with $D =$ 0.41 day^{-1} to obtain μ_{app}; i is calculated using Equation. The apparent and true specific growth rates are shown in Table 8.1 and plotted in Fig. 8.2b. Note that μ_{app} is essentially constant during the increase in cell concentration, while μ reaches a maximum and then continually declines, as the viability increases, during this period.

Batch culture

In this case $F_i = F_o = F = 0$. Reactor samples (i.e. $F_o \neq 0$) do not affect the analysis because they reduce the culture volume without altering any of the concentrations. However, acid or base additions do affect concentrations. If the volume added is significant the reactor effectively becomes a fed-batch reactor. In this case Equations would apply. For negligible cell lysis the expressions for μ_{app}, μ and k_d for batch culture can be obtained from Equations by setting $D = 0$:

$$\mu_{app} = \frac{d\ln(n)}{dt}; \quad \mu = \frac{\mu_{app}}{f_v}; \quad k_d = \mu - \frac{d\ln(n_v)}{dt}$$

TABLE 8.1 DATA FOR THE GROWTH RATE EXAMPLE CALCULATIONS ILLUSTRATED

Culture time (days)	In (total cells ml^{-1})	Fraction viable	Slope (day^{-1})	μ_{app} (day^{-1})	μ (day^{-1})
−0.88	14.14	0.62	0.00	0.41	0.66
−0.24	14.15	0.63	0.00	0.41	0.65
0.26	1.4.16	0.68	0.00	0.41	0.60
0.78	14.07	0.68	0.00	0.41	0.60
1.12	14.16	0.69	0.08	0.49	0.72
1.74	14.23	0.75	0.17	0.58	0.77
2.12	14.34	0.72	0.17	0.58	0.81
2.73	14.40	0.76	0.17	0.58	0.76
3.11	14.52	0.78	0.17	0.58	0.74
3.73	14.56	0.80	0.17	0.58	0.72
4.12	14.64	0.81	0.17	0.58	0.72
4.72	14.76	0.82	0.17	0.58	0.71
5.11	14.84	0.85	0.17	0.58	0.68
5.74	14.95	0.82	0.00	0.41	0.50
6.12	14.84	0.81	n.d.	n.d.	n.d.
6.78	14.85	0.75	−0.05	0.36	0.48
7.22	14.87	0.69	0.00	0.41	0.59
7.80	14.85	0.69	0.00	0.41	0.59
8.10	14.83	0.68	0.00	0.41	0.60
8.73	14.89	0.71	0.00	0.41	0.58
9.12	14.86	0.71	0.00	0.41	0.58
9.72	14.86	0.70	0.00	0.41	0.59

Equation is valid for small k_l, which is normally the case until the culture reaches the late stationary or death phase.

Semi-batch culture

For semi-batch culture $F_o = 0$, but $F_i \neq 0$ so V is not constant. In this case it is easiest to solve Equations in terms of the number of cells (nV or n_vV, respectively) in the vessel, as opposed to the cell concentrations:

$$\mu_{app} = \frac{d\ln(nV)}{dt}; \quad \mu_{app} = \frac{d\ln(n_vV)}{dt}$$

where $V(t) = \int F_i \, dt$. For small k_l, we again have $\mu = \mu_{app}/f_v$. From Equations is obtained:

$$k_d = \mu - \frac{d\ln(n_vV)}{dt}$$

Note that these relations can be used for periodic, as well as continuous, medium additions (i.e. for $F_i(t)$ a discrete or continuous function). Equations neglect sampling ($F_o \neq 0$). In this case reactor

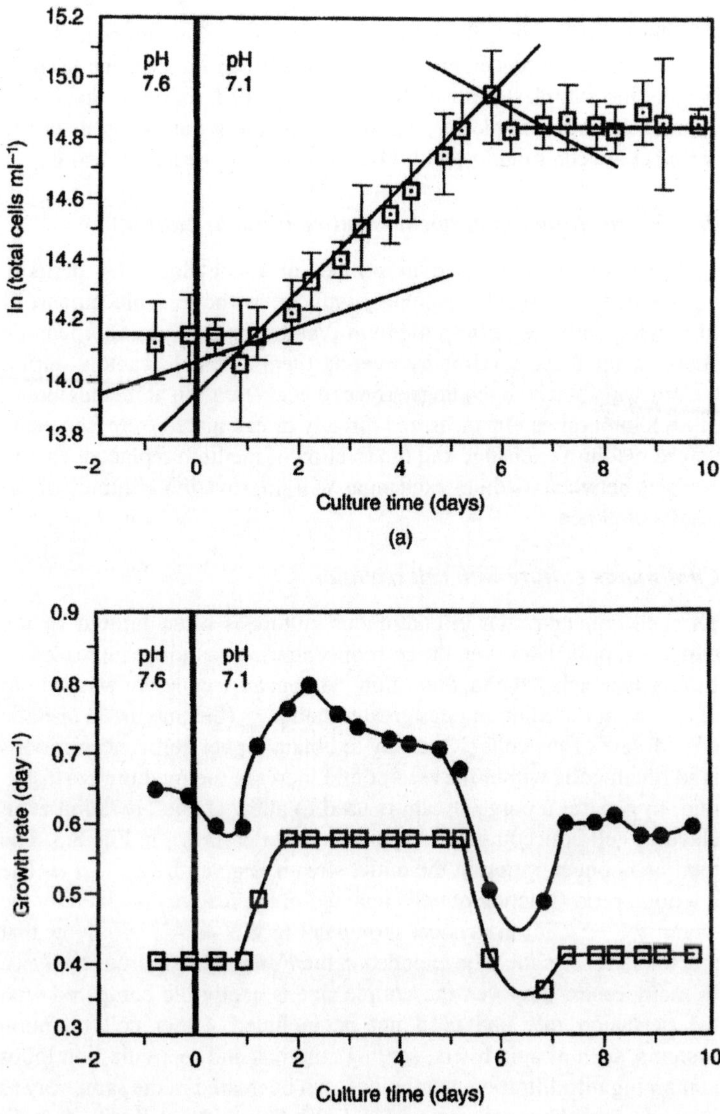

Fig. 8.2 A pH step change continuous culture experiment. The reactor pH was decreased from 7.6 to 7.1 at time zero. The dilution rate was maintained at 0,41 day"¹ throughout the experiment, (a) Natural logarithm of total cell concentration versus culture time. Tangent lines are shown for slope determination. Error bars represent the standard deviation based on four replicate cell counts, (b) Apparent (μ_{app}, □) and true (μ, ●) specific growth rates calculated from (a) and Equations 4.2.7 and 4.2.8.

samples do affect the culture because they reduce the volume and alter cell and nutrient concentrations. However, Equations can be used as long as the samples removed represent a small fraction of the vessel volume. If large samples are removed, Equations can be used for the periods between samples, with $V(t)$ corrected for the next interval.

Periodic medium replacement (semi-continuous culture)

Cells are often maintained in spinner or shake flasks by periodic replacement of spent medium (with or without replacement of suspended cells) with fresh medium. Values for μ_{app}, μ and k_d can be obtained for these systems by treating them as batch reactors during the intervals between feedings. The cell concentration at the beginning of each interval can be measured directly or calculated from, the value before medium exchange and the fraction of medium replaced. Taking samples between medium exchanges will improve the accuracy of the results obtained.

Continuous culture with cell retention

The cell concentration in continuous culture is often limited by the nutrient supply. However, the cell concentration cannot be increased by simply increasing the medium flow rate because cells are washed out of the reactor at dilution rates greater than μ_{max} (the maximum specific growth rate of the cells). One way to obtain higher cell concentrations is to retain cells within the reactor and increase the medium perfusion rate. In general a purge stream is used to allow stable operation at an elevated cell concentration. A typical system is shown in Fig. 8.3. The total cell concentration in the outlet stream is given by $n_0 = \alpha_s n$. The retention ratio (fraction of cells retained in the reactor) is given by the quantity $1-\alpha_s$. For the system shown in Fig. 8.3, $\alpha_s = F_2/F_o$. Note that reactor samples should be included in the F_2 term for rigorous analysis. In many cases, however, the sample size is negligible compared with the perfusion rate and need not be included. Other cell retention systems, such as spin filters, settling columns and recircula-tion loops containing ultrafiltration membranes, can be treated in the same way as long as α_s can be measured or calculated. For constant V with $F_i = F_o = F$, the total cell balance around this system becomes:

$$\frac{dn}{dt} = -\alpha_s nD + \mu_{app} n$$

Expressions for μ_{app}, μ and k_d can be obtained by replacing D with $\alpha_s D$ in Equations:

$$\mu_{app} = \frac{d\ln(n)}{dt} + \alpha_s D; \mu = \frac{\mu_{app}}{f_v}$$

$$k_d = \mu - \alpha_s D - \frac{d\ln(n_v)}{dt}$$

At steady state $\mu_{app} = \alpha_s D < D$. It should be noted that these equations assume equal α_s values for viable and non-viable cells. This will be true for separation devices such as filters, which do not allow any cells to pass (e.g. Fig. 8.3), but may not be true for settling columns, which generally have greater retention (i.e. smaller α_s) of viable cells. In the latter case, the investigator should measure α_s for viable and non-viable cells by taking cell counts in the reactor and at the settling column outlet. If there is no significant difference in α_s for viable and non-viable cells or if the reactor viability is very high, then the above equations may be used. Otherwise the $\alpha_s n$ term in Equation must be replaced by ($\alpha_{sv} n_v + \alpha_{sd} n_d$) where $1 - \alpha_{sv}$ and $1 - \alpha_{sd}$ represent the retention ratios for viable and non-viable cells, respectively.

Cases with significant cell, lysis

Cell lysis generally becomes important when viability is low. Examples include continuous culture (with or without cell retention) at low dilution rates and the late stationary and death phases of batch culture. Cell lysis is also important in stirred reactors with very high agitation rates. In order to quantify cell growth parameters under these conditions, the lysed cells must be accounted for. One way to do this is to measure the amount of the cytosolic enzyme lactate dehy-drogenase (LDH) released into the medium. A procedure for measuring LDH activity is described.

Lactate dehydrogenase is a useful marker because it is released upon cell death and is stable over short periods of time (5% loss per day) so that the concentration of LDH in the medium provides an estimate of the total number of (intact plus lysed) dead cells. Equations are still valid, but a balance needs to be included for the effective cell concentration n_e, which is equal to the total cell concentration plus the concentration of cells that have lysed (i.e. what the total cell concentration would be if no cells had lysed):

$$\frac{d(n_e V)}{dt} = 0 - n_e F_o + \mu_{app}^{e} n_e V$$

where μ_{app}^{e} is the effective cell-derived apparent specific growth rate. Because new cell mass is only produced by viable cells, we have:

$$\mu_{app}^{e} n_e = \mu n_v$$

The effective cell concentration is given by:

$$n_e = n_v + \frac{C_{LDH}}{\gamma_{LDH}}$$

where C_{LDH} is the concentration of LDH in the culture medium and γ_{LDH} is the LDH content per cell; γ_{LDH} has been shown to be relatively constant in viable cells. In most suspension cultures γ_{LDH} is zero in dead cells (i.e. all LDH is released upon cell death). However it has been shown that dead cells (i.e. cells that do not exclude Trypan blue) detached from microcarriers have a γ_{LDH} value approximately 50% of that for attached viable cells. The dead cell LDH content should be verified for each system and γ_{LDH} can be obtained by lysing a known number of viable cells and measuring the amount of LDH released. If dead cells are found to contain LDH, then Equation must be modified to give:

$$n_e = n_v + \frac{C_{LDH}}{\gamma_{LDH}} + \beta_d n_d$$

where β_d is the fraction of γ_{LDH} retained by dead cells, so the amount of LDH released by dead cells is equal to $(1 - \beta_d)\gamma_{LDH}$. One way to test for the importance of cell lysis is to compare the value of C_{LDH}/γ_{LDH} with that of n_d in the culture. Significant lysis is evidenced by n_a being less than C_{LDH}/γ_{LDH} (or n_d being less than $C_{LDH}/\gamma_{LDH} + \beta_d n_d$ if dead cells retain some LDH).

In a short-term batch culture or in a continuously perfused system with low medium residence time, the degradation of LDH may not be significant and may therefore be ignored. However, if analysis is continued over several days at low viabilities or high lysis rates, LDH degradation will be significant. Therefore, the calculated effective cell concentration will be underestimated if no correction is made. The simplest modification is to increase the first day's LDH value by 5% to obtain a corrected C_{LDH} for that day. Five per cent of this value is then added to the C_{LDH} value for the next day, and the process is repeated to calculate C_{LDH} for the second day, and so on. The more rigorous approach of accounting for degradation by writing an unsteady-state mass balance on LDH and solving for an effective C_{LDH} (the concentration if no degradation occurred) at each sample point may be justified for extended experiments. An average degradation rate of 5% per day may be assumed if no data are available. However, the degradation rate may vary with culture parameters such as pH and even with cell number and viability (due to release of proteolytic enzymes). The degradation rate in a particular system can be measured by removing cells from a sample and following the LDH concentration over time.

1. *Continuous culture.* For a constant-volume vessel Equation can be rearranged to give:

$$\mu_{app}^{e} = \frac{d\ln(n_e)}{dt} + D$$

where n_e is obtained as described above. The specific growth rate is given by:

$$\mu = \mu_{app}^{e}\left(\frac{n_e}{n_v}\right)$$

The specific death rate k_d is given by Equation while k_l can be obtained from Equations:

$$k_l = \mu\left(\frac{n_v}{n_d}\right) - \left(D + \frac{d\ln(n)}{dt}\right)\left(\frac{n}{n_d}\right)$$

2. *Batch culture.* In this case the importance of cell lysis can be estimated via the extent of deviation from a straight line in a plot of C_{LDH} versus n_d. As for the case without significant cell lysis, the batch parameters can be obtained by setting $D = 0$ in Equations:

$$\mu_{app}^{e} = \frac{d\ln(n_e)}{dt}; \mu^{e} = \mu_{app}^{e}\left(\frac{n_e}{n_v}\right);$$

$$k_d = \mu - \frac{d\ln(n_v)}{dt}; k_l = \mu\left(\frac{n_v}{n_d}\right) - \left(\frac{d\ln(n)}{dt}\right)\left(\frac{n}{n_d}\right)$$

Cells on microcarriers

Cells on microcarriers are normally cultured in batch or perfusion reactors. In perfusion systems microcarriers are normally retained, but free cells exit with the outlet stream. The balance on total cells in a constant-volume perfusion system is given by:

$$\frac{dn}{dt} = -n_F D + \mu_{app} n$$

with $n = n_A + n_F$, where n_A is the concentration of attached cells and n_F is the concentration of free cells. If attached cells are present predominantly as a mono-layer on the bead, then it is generally found that all of the attached cells are viable. In this case $n_v = n_A + f_{Fv} n_F$ (where f_{Fv} is the viable fraction of free cells) and:

$$\mu_{app} n = \mu_A n_A + \mu_F (f_{Fv} n_F)$$

or:

$$\mu_A = \mu_{app}\left(\frac{n}{n_A}\right) - \mu_F f_{Fv}\left(\frac{n_F}{n_A}\right)$$

where μ_{app} is given by Equation as:

$$\mu_{app} = \frac{d\ln(n)}{dt} + D\left(\frac{n_F}{n}\right)$$

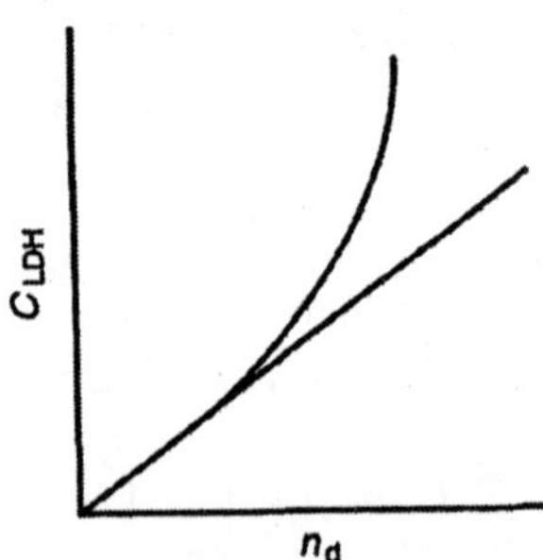

Fig. 8.4 Schematic of medium LDH concentration (C_{LDH}) versus dead cell concentration (n_d) in batch culture. For systems with minimal cell lysis, the relationship is a straight line with slope equal to the LDH content per cell (γ_{LDH}). The curve with $C_{LDH} > n_d$ indicates significant dead cell lysis, with greater deviation from the straight line indicating more extensive lysis.

Because free cells are removed with the outlet medium, n_F is normally much less than n. However, the second term in the RHS of Equation may be important because the increase in cell concentration often slows as cells reach confluence on the beads. Parameter μ_A can be calculated from μ_{app} if μ_F is known, and μ_F can be determined by allowing the microcarriers to settle, growing the remaining suspended cells in batch culture and analysing the data as described above. Because $n_F \ll n$, it should be possible to obtain an accurate estimate for μ_F at the conditions of interest before the nutrient supply in the medium removed from the reactor is depleted. For many cell types μ_F is very small for single cells and f_{Fv} is of the order 0.5. In such cases, the second term in Equation can be neglected, especially since $n_F \ll n$.

The balance on adherent cells is given by:

$$\frac{dn_A}{dt} = \mu_{app}^{A} n_A = \mu_A n_A - k_R n_A$$

where μ_{app}^{A} is the apparent specific growth rate of adherent cells and k_R is the rate of cell release from the beads; μ_{app}^{A} can be used directly to characterize the process or it can be used with μ_A to determine k_R, as in:

$$k_R = \mu_A - \frac{d\ln(n_A)}{dt}$$

Although adherent cells are not removed with the outlet medium, they are removed with reactor samples. If large samples are removed for cell characterization, then the cell concentration should be adjusted for use in calculating the various specific growth rates.

For batch microcarrier culture, Equations apply with $D = 0$. However, the contribution from free cells in Equations will be greater because free cells are not removed from the system. High n_F values may lead to aggregate formation, which further complicates analysis.

Cell dumps and spheroids

The situation is much more complicated for multilayers of cells on microcarriers or for cells that grow in clumps or spheroids. The specific growth rate in such systems is normally a function of cell location. For example, cells at the outside of tumour cell spheroids generally proliferate rapidly, while those further in are quiescent. For spheroids with a radius larger than about ten cell diameters, there is normally a necrotic core of dead cells. In these complex systems it is only possible to determine apparent growth rates. These can be expressed in terms of total cell numbers, viable cell numbers or spheroid volume. Another difficulty with these systems is that accurate cell counts are hindered by cell clumps that resist dissociation with trypsin. One way around this is to lyse the cells with surfactant and count the nuclei released. However, analysis is complicated because the nuclei from recently dead cells are preserved, while those from cells dead for long periods are lost.

Immobilized-cell reactors

Cell density is rarely measured directly in immobilized-cell reactors because it is not possible to obtain a single-cell suspension suitable for counting without disturbing the reactor. Specialized techniques are available for indirect non-invasive determination of cell density, but most researchers estimate cell density via nutrient consumption rates. Methods for estimation of cell density from nutrient consumption rates and an analysis of the reliability of such estimates are discussed below.

Cell Metabolism

The best way to characterize cell metabolic patterns is in terms of consumption rates and production rates per viable cell (collectively termed metabolic quotients). This allows much easier comparison between different cell types or culture conditions than do plots of changes in the levels of nutrients or products over time. It also shows directly how metabolic patterns change with time during batch growth or in response to changes in culture conditions for a given experiment. As discussed below, metabolic quotients can be readily calculated from metabolite and viable cell concentration data.

Metabolic quotients in stirred vessels

The equations in this section are applicable to any system in which the viable cell density can be determined. This includes all of the systems described above except for immobilized cell reactors.

Dissolved substrates and products

For stirred vessels such as those shown in Figs 8.1 and 8.3, a balance on the concentration of a dissolved metabolite M gives:

$$\frac{d(MV)}{dt} = F_i M_F - F_o M + q_M n_v V$$

where M_F is the concentration in the feed stream. M is the concentration in the vessel and the metabolic quotient q_M is the net specific formation rate of M (mol M time^{-1} viable cell $^{-1}$). Equation applies for reactors with cell retention because small metabolites are not retained along with the cells. However, the relation must be modified if product proteins are (partially) retained in the vessel.

For constant-volume vessels we obtain:

$$q_M = \frac{D(M - M_F) + (dM/dt)}{n_v}$$

Note that q_M will be negative for nutrients that are consumed by the cells and positive for products. In order to obtain positive values for all of the metabolic quotients, it is customary to speak in terms of the specific substrate (nutrient) consumption rate q_s and the specific product formation rate q_P

$$q_S = \frac{D(S_F - S) - (dS/dt)}{n_v}; \quad q_P = \frac{D(P - P_F) + (dP/dt)}{n_v}$$

For batch systems, Equation becomes:

$$q_S = \frac{-(dS/dt)}{n_v}; \quad q_P = \frac{(dP/dt)}{n_v}$$

For semi-batch systems the metabolic quotients are as in Equation:

$$q_S = \frac{F_i S_F - [d(SV)/dt]}{n_v V}; \quad q_P = \frac{-F_i P_F + [d(PV)/dt]}{n_v V}$$

Note that V (and possibly F_i) varies with time.

Metabolic quotients are easily determined in continuous culture. At steady state, the derivative terms in Equation may be set equal to zero, thereby allowing easy calculation of the specific consumption and production rates. To calculate the metabolic quotients during transients (in continuous, batch or semi-batch culture), it is necessary to obtain dS/dt and dP/dt (or d(SV)/dt and d(PV)/dt) for each time point. Scatter in S and P data will result in large fluctuations in q_S and q_P if the derivatives are determined by connecting points from successive samples. These fluctuations may mask trends in the data, so it is recommended that the derivatives be obtained using the slopes of smoothed curves through the S and P data points. Optionally, one could smoothe out scatter in the data by fitting the metabolite and product concentration versus time plots with low-order polynomial functions and then obtaining the derivatives analytically. Regardless of the method used to obtain the derivative, the measured values of S, P and n_v should be used for the other terms in Equations.

Example 2

Table 8.2 contains metabolite and cell concentration data for a continuous culture experiment with $D = 0.54$ day^{-1} in which a glucose step change was implemented. The initial steady state was established at a feed glucose concentration of 5.2 mM, while the residual glucose concentration in the reactor was close to 0 mM. At time zero, the glucose concentration in the reactor was increased to 8.3 mM and the feed concentration was increased to 13.8 mM. The viable cell concentration increased from an initial steady-state value of ~1.6×10^6 cells ml^{-1} to a final steady-state value of ~2.5×10^6 cells ml^{-1} (Table 8.2). Fig. 8.5 shows the glucose and lactate concentration profiles. Derivatives for the lactate concentration versus time plot were determined from the slopes of the lines shown in Fig. 8.5b, and are shown in Table 8.2. Derivatives for the glucose concentration versus time plot were determined from the plots in Fig. 8.6, and are shown in Table 8.2. At the initial and final steady states, the slope is taken to be zero (Figs. 8.5a and 8.6c). During the first day after the step change, the slope is essentially constant, as shown in Fig. 8.6a. The slope at the transition time of 1.44 days is determined by fitting the region near that point with a third-order polynomial (Fig. 8.6b) and analytically obtaining the derivative. The specific glucose consumption and lactate production rates calculated from Equation are presented in Figs. 8.7a and 8.7b, respectively. Note that q_{glc} and q_{lac} are very sensitive to the reactor glucose concentration.

Estimation of metabolic quotients in flask cultures

Consumption and production rates are often measured by incubating cells in media and measuring the substrate or product concentration at the beginning and end of a specified period of time. Equation becomes:

$$q_S = \frac{-(\Delta S / \Delta t)}{n_{Vave}}; \quad q_P = \frac{(\Delta P / \Delta t)}{n_{Vave}}$$

where n_{Vave} is the average viable cell concentration. ΔS and ΔP represent the difference between the initial and final substrate and product concentrations, respectively, and Δt represents the time between samples. It should be noted that metabolic quotients obtained over long periods reflect average values over the interval. The method used to calculate n_{Vave} depends upon the amount of cell growth that occurs during the time period Δt. For very long incubation times (Δt much greater than the doubling time), multiple samples should be taken. For moderate times during which significant growth occurs, one may assume a linear dependence of cell concentration on time. In this case n_{Vave} is simply the arithmetic average of the initial and final cell concentrations. For non-confluent adherent cell cultures, the initial cell concentration may be estimated using plating efficiencies obtained from experiments in replicate flasks. For confluent monolayers with diminished cell growth rates, it may be acceptable to use the final cell count for n_{Vave}.

Oxygen

The use of oxygen in cell culture is characterized by the specific oxygen consumption rate. Cultures grown in incubators are generally oxygenated by simple diffusion. Oxygen may be provided to cells in bioreactors via perfusion with oxygenated medium or via transport from the headspace, gas bubbles and/or semi-permeable tubing. In general:

$$q_{O_2} = \frac{D(C_{O_{2F}} - C_{O_2}) + K_L a[(P_{O_2} / H) - C_{O_2}] - [d(C_{O_2}) / dt]}{n_v}$$

where C_{O2} and C_{O2F} are the dissolved oxygen concentrations (mM) in the vessel and feed stream, respectively, P_{O2} is the oxygen partial pressure (mm Hg) in the gas phase, H (~ 760 mm Hg mM^{-1}) is the Henry's law constant for oxygen in culture medium at 37°C and $K_L a$ is the volumetric oxygen mass transfer coefficient. In most agitated vessels, the contribution from oxygen in the feed stream is negligible. This yields:

TABLE 8.2 DATA FOR THE GLUCOSE STEP CHANGE EXAMPLE IFLUSTRATE

Time (days)	glc (mM)	lac (mM)	wv (cells (ml^{-1} ×10^{-6})	glc slope (mM day^{-1})	lac slope (mM day^{-1})	q_{glc} (× 10^{-9})	q_{lac} (× 10^{-9})	$Y'_{lac,glc}$	q_{ATP} ×10 (P/O = 3)	q_{ATP} ×10 (P/O = 3)
−1.77	0.06	10.80	1.69	0.00	0.00	1.64	2.62	1.59		
−1.15	0.11	10.56	1.67	0.00	0.00	1.65	2.57	1.56	2.30	3.32
−0.76	0.06	10.01	1.65	0.00	0.00	1.68	2.43	1.44	2.36	3.42
−0.14	0.00	11.20	1.56	0.00	0.00	1.80	2.98	1.65	2.32	3.33
0.00	8.28	10.47	n.d.	−5.93	6.67	n.d.	n.d.	n.d.	n.d.	n.d.
0.04	8.17	10.90	n.d.	−5.93	13.33	n.d.	n.d.	n.d.	n.d.	n.d.
0.09	7.89	11.31	1.63	−5.93	13.33	5.60	11.1	1.98	2.81	3.66
0.17	7.33	12.42	1.59	−5.93	13.33	5.93	11.7	1.98	2.91	3.78
0.25	6.83	14.27	1.51	−5.93	13.33	6.42	13.0	2.03	3.03	3.89
0.34	6.44	15.28	1.57	−5.93	13.33	6.31	12.9	2.04	3.09	3.99
0.51	5.33	16.90	1.75	−5.93	13.33	6.00	12.0	2.00	2.97	3.85
0.94	2.78	22.36	2.15	−5.93	13.33	5.53	11.2	2.02	2.82	3.67
1.44	0.44	29.56	2.21	−1.69	0.00	4.03	6.59	1.64	2.37	3.23
1.85	0.28	27.66	2.40	0.00	−3.30	3.04	4.26	1.40	2.14	3.00
2.27	0.33	27.22	2.64	0.00	−1.10	2.75	4.62	1.68	2.22	3.10
2.85	0.39	26.89	2.47	0.00	0.00	2.93	5.31	1.81	2.31	3.20
3.24	0.33	26.00	2.45	0.00	0.00	2.97	5.16	1.74	n.d.	n.d.
3.85	0.28	30.11	2.38	0.00	0.00	3.07	6.24	2.03	2.42	3.32
4.23	0.28	26.33	2.56	0.00	0.00	2.85	5.01	1.76	2.36	3.30
4.85	0.28	27.22	2.43	0.00	0.00	3.00	5.47	1.82	2.39	3.31
5.27	0.28	26.56	2.55	0.00	0.00	2.86	5.07	1.77	2.20	3.05
5.85	0.22	26.44	2.52	0.00	0.00	2.91	5.11	1.76	2.41	3.37

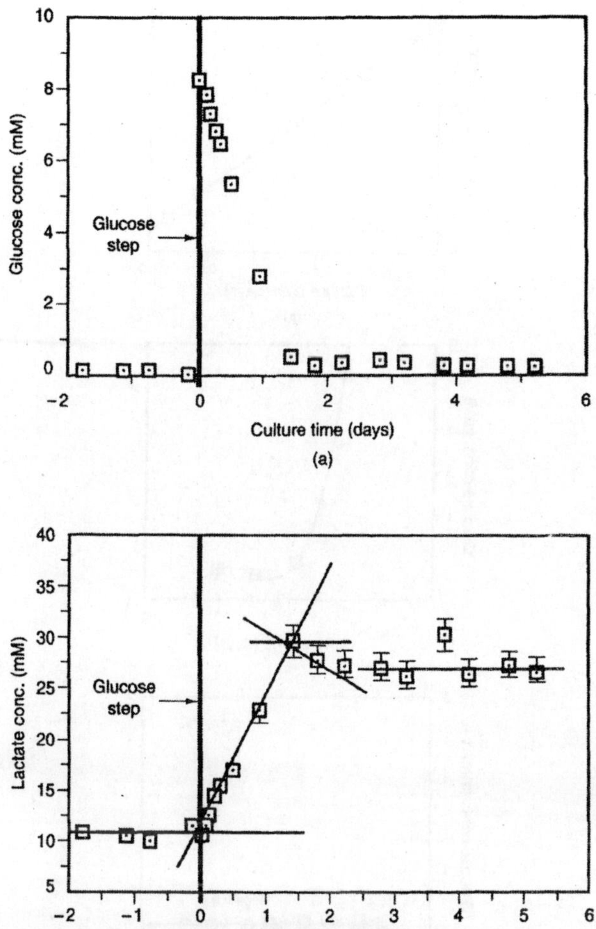

Fig. 8.4 Glucose step change continuous culture experiment. Initial feed glucose concentration was 5.2 mM. At time zero the feed glucose concentration was increased to 13.8 mM and the reactor glucose concentration was increased to 8.3 mM. The reactor dilution rate was maintained at 0.54 day^{-1} throughout the experiment, (a) Glucose concentration (mM) as a function of culture time. The estimated error is ± 0.02 g l^{-1} based on the variability of a 2 g l^{-1} glucose standard and phosphate-buffered saline (PBS) control, (b) Lactate concentration (mM) as a function of culture time. Tangent lines are shown for slope determination. Error bars represent the standard deviation based on a standard curve.

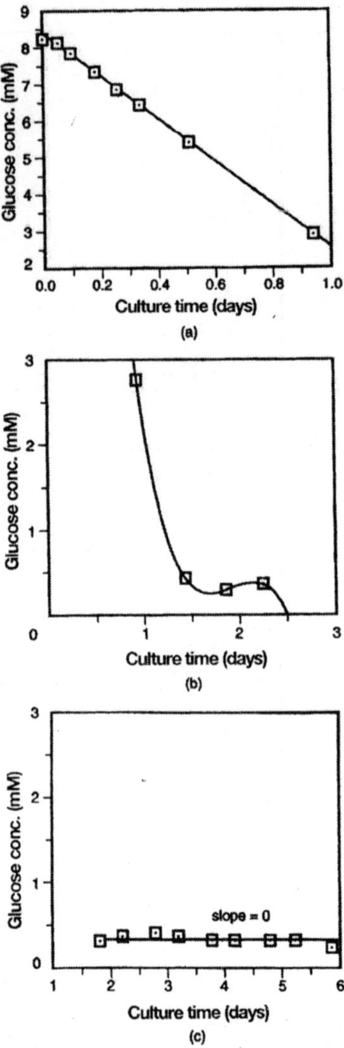

Fig. 8.5 Plots for determination of the derivative of glucose concentration as a function of time for the glucose step change experiment described in Fig. 8.5. The slope is zero for times less than zero (initial steady state), (a) Linear fit for the first day after the step change. Least-squares analysis gives a slope of −5.93 mM day^{-1}. (b) Polynomial fit to obtain the derivative at the transition point (1.44 days). A third-order fit gives [glucose] = 21.28 $-33.65t+17.76t^2-3.08t^3$. The derivative is then d[glucose]/dt = $-33.65 + 35.52t - 9.25t^2$. For $t = 1.44$ days, the slope is -1.69 mM day^{-1}, (c) Zero slope for all points after 1.44 days (final steady state).

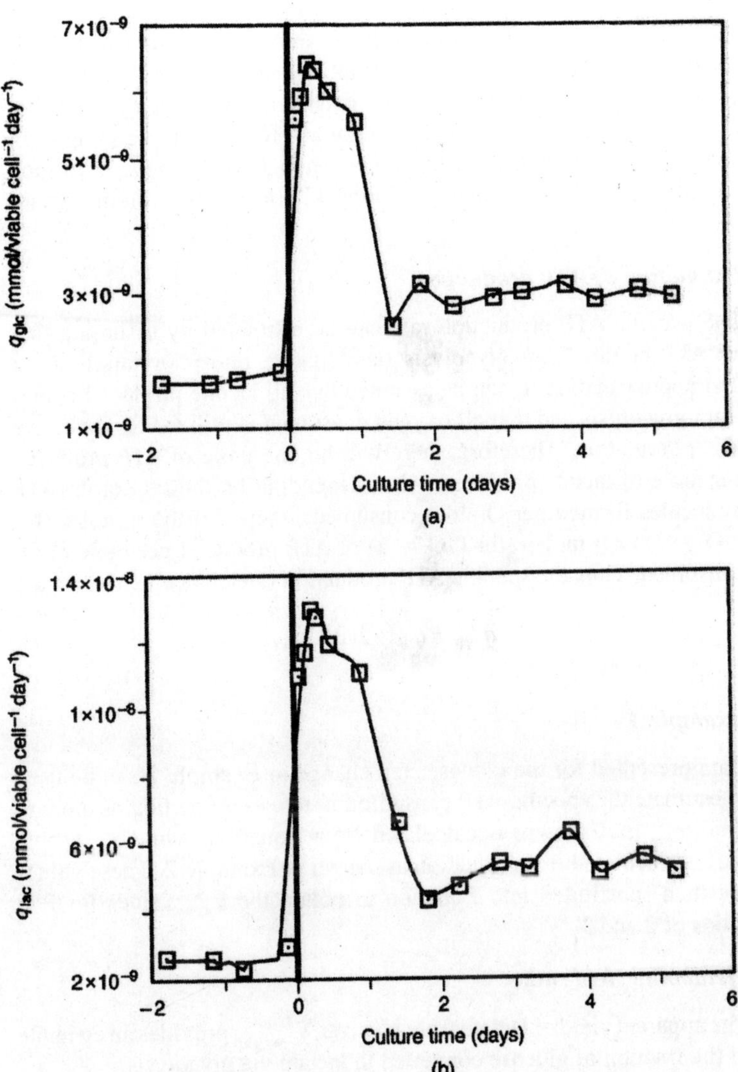

Fig. 8.6 Metabolic quotients for the glucose step change experiment described in Fig. 8.5. (a) Specific glucose consumption rate (mmol glucose consumed viable cell^{-1} day^{-1}) as a function of culture time, (b) Specific lactate production rate (mmol lactate produced viable cell^{-1} day^{-1}) as a function of culture time.

$$q_{O_2} = \frac{K_L a[(P_{O_2}/H) - C_{O_2}] - [d(C_{O_2})/dt]}{n_v}$$

The derivative is evaluated according to the methods described above for other metabolites. As before, point values should be used for all of the other terms. Procedures for determination of $K_L a$. Note that $K_L a$ is assumed to be constant throughout. However, large samples in batch culture could deplete the volume enough to change $K_L a$. In such circumstances, a functionality relating $K_L a$ to reactor volume would be useful.

Estimation of ATP production

The specific ATP production rate can be estimated by assuming that all ATP comes from glycolysis or oxidative phosphorylation. As a first approximation, it can be assumed that all lactate produced comes from glycolysis and that all oxygen consumed is utilized for oxidative phosphorylation. Therefore, there will be one mole of ATP produced per mole of lactate produced. Also, depending on the P/O ratio (ATP molecules formed per O atom consumed), there will be 6 moles (for P/O = 3) or 4 moles (for P/O = 2) of ATP produced per mole of O_2 consumed. Thus the specific ATP production rate can be expressed as:

$$q_{ATP} = q_{lac} + 2(P/O)q_{O_2}$$

Example 3

Data presented for the glucose step change in Example 2 can be used to estimate the specific ATP consumption rate as a function of the P/O ratio; q_{O2} (not shown) is calculated from Equation, while q_{lac} (Table 8.2) is calculated from Equation as shown in Example 2. These values are then substituted into Equation to obtain the q_{ATP} values for P/O ratios of 2 and 3.

Metabolite yield ratios

The apparent yield of lactate from glucose, $Y'_{lac,glc}$, provides an estimate of the fraction of glucose converted to lactate via glycolysis:

$$Y'_{lac.glc} = \frac{q_{lac}}{q_{glc}}$$

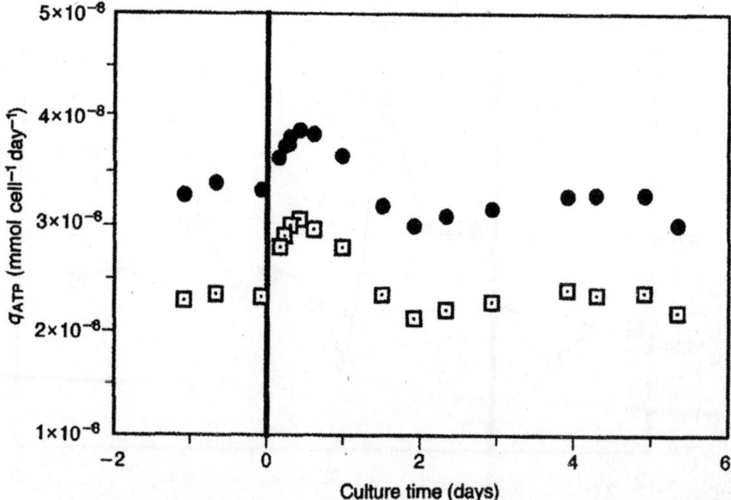

Fig. 8.7 Specific ATP production rate as a function of culture time for P/O ratios of 3 (●) and 2 (□) for the glucose step change experiment described in Fig. 8.5.

This is an apparent yield because lactate can be produced via metabolism of other substrates such as glutamine, and because pyruvate derived from glycolysis can be converted into other compounds such as alanine. The theoretical maximum yield is 2 because no more than two molecules of lactate can be obtained from a single molecule of glucose. However, production of lactate from glutamine may result in yields greater than 2. The true yield of lactate from glucose, $Y_{lac,glc}$, would be the number of molecules of lactate directly obtained from the metabolism of glucose. True yields must be determined by metabolic labelling studies. For example, the carbon atoms in glucose could be radiolabelled so that any radioactive lactate molecules must have been produced from glucose. Nuclear magnetic resonance techniques with ^{13}C-labelled glucose can also be used.

Apparent yields are useful for characterizing and following changes in cell metabolism because changes in the apparent yield indicate changes in the use of alternative metabolic pathways. Typically, an apparent yield ratio relates the formation of a product to the consumption of a substrate. However, a yield ratio can also relate two substrates, as in the case of oxygen and glucose: $Y'_{O2,glc}$ gives an estimate of the extent of oxidative metabolism relative to glycolysis. Other useful yield ratios

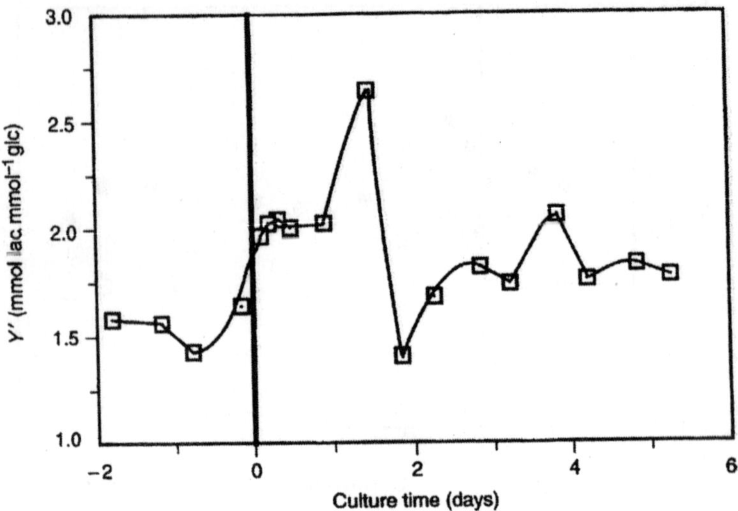

Fig. 8.8 Apparent yield of lactate from glucose (mmol lactate produced per mmol[1] glucose consumed) as a function of culture time for the glucose step change experiment described in Fig. 8.5.

include oxygen/glutamine, ammonia/glutamine and product/oxygen. Because the cell concentrations in Equation cancel, there is often less scatter in apparent yield data than in the individual consumption rates.

Example 4

The apparent yield of lactate from glucose is easily calculated for the glucose step change described in Example 2. Both q_{glc} and q_{lac} are shown in Table 8.2. From Equation, the apparent yield is obtained by dividing the specific lactate production rate by the specific glucose consumption rate. The resulting $Y'_{lac,glc}$ values are plotted as a function of time are listed in Table 8.2. Note that the fraction of glucose converted to lactate also increases at higher glucose concentrations.

Calculation of apparent yields without n_v

An apparent yield is typically calculated by dividing a specific production rate by a specific consumption rate. As shown in Equations, the viable cell concentration terms cancel out. Therefore, the apparent yield may be calculated by evaluating the numerators in Equation for product and substrate and taking their ratio. In batch culture, the analysis is even simpler. An apparent yield ratio based on Equation:

$$Y'_{P,S} = \frac{q_P}{q_S} = \frac{dP}{-dS}$$

The yield is easily calculated by evaluating the derivatives of product concentration versus time and substrate concentration versus time and taking their ratio. Alternatively, product concentration is plotted against substrate concentration and the absolute value of the slope at any point is equal to the apparent yield. The important point to note here is that apparent yield ratios may be obtained even if specific consumption and production rates cannot be calculated because viable cell data are unavailable. For cases in which only endpoint data are available, an average apparent yield can be calculated. In this case, the apparent yield is simply the ratio of the change in the concentration of product to the change in the concentration of substrate:

$$Y'_{P,S} = \frac{\Delta P}{-\Delta S}$$

Cell yield on substrate

Using the concept of yield ratios, we can define a quantity relating cell growth to substrate utilization:

$$Y'_{n,S} = \frac{\mu}{q_S}$$

where μ is the specific growth rate and q_s is the specific consumption rate for substrate S. Yield parameter $Y'_{n,S}$ is a sensitive parameter that varies with cell growth rate and culture conditions, especially the relative amounts of different substrates, and is useful for detecting changes in substrate utilization as well as for identifying limiting nutrients in media development.

ATP maintenance energy model

The maintenance energy model states that energy consumption can be divided into that needed for growth and that needed for maintenance. Through the use of yield ratios, an empirical expression can be written for a specific substrate consumption rate as a function of the maintenance energy requirement and the specific production rates of all products that require the consumption of that substrate. This type of model is useful in microbial systems, for which there is frequently only one carbon source. However, such a model is not very useful for animal cell culture because multiple substrates are required and the different products can be synthesized from alternative substrates.

A more useful model is the ATP maintenance energy model. The specific utilization rate of ATP can be expressed as:

$$q_{\text{ATP}} = \frac{\mu}{Y'_{\text{n,ATP}}} = \frac{\mu}{Y_{\text{n,ATP}}} + m_{\text{ATP}}$$

The advantage of Equation is that all substrates provide energy via ATP so it does not matter what nutrients are used as long as q_{ATP} can be estimated. In Equation energy for product formation would be included in the $Y'_{\text{n,ATP}}$ term for growth-associated products and in the m_{ATP} term for non-growth-associated products. Alternatively, a specific product term could be added to Equation in the form of $q_{\text{p}}/Y'_{\text{P,ATP}}$ estimated m_{ATP} and $Y'_{\text{n,ATP}}$ for hybridoma cells using Equation and the information shown in Examples 2 and 3.

Immobilized-cell reactors

Estimation of cell density using metabolic quotients

It is often impossible to determine experimentally the cell concentration in immobilized-cell reactors. However, it is usually very easy to measure oxygen and metabolite concentrations in the inlet and outlet streams. From the metabolite concentration versus time data, a volumetric (per unit reactor volume rather than per cell) production rate can be determined:

$$Q_{\text{M}} = D(M_{\text{F}} - M_{\text{o}})$$

Here, M_{o} is the metabolite concentration in the outlet stream and M_{F} is the concentration in the feed stream to the reactor (M is often oxygen or glucose). If oxygen is also supplied via transfer from gas-filled hollow fibres, an extra term must be added to Equation Terms Q_{M} and q_{M} are related by:

$$q_{\text{M}} = \frac{Q_{\text{M}}}{n_{\text{v}}}$$

If q_M is obtained from experiments in stirred suspension or microcarrier cultures, as described above, then, it may be possible to estimate n_v from $Q_{M/qM}$. However, care must be exercised because q_M can be altered by changes in the culture environment. For example, q_M for cells immobilized in agarose beads or hollow fibres may be different from q_M for the same cells grown in the same medium in a stirred suspension reactor. In addition, q_M may change over time due to changes in cell, nutrient and byproduct concentrations. Analysis of cell density and immobilization effects is complicated by the presence of nutrient concentration gradients. However, stirred vessels with cells immobilized in spherical beads may be used to estimate these effects if the concentrations of glucose, oxygen and other nutrients are high enough to ensure that the concentrations at the centre of the beads are in the plateau regime (i.e. metabolic quotients independent of nutrient concentrations).

Because apparent metabolic yields can be determined without knowledge of n_v, Y' values can be monitored to detect changes in metabolic patterns of cells in immobilized-cell bioreactors. For example, $Y'_{lac,glc}$ is easy to obtain and provides a good measure of the fate of the primary nutrient glucose. Relatively constant values for $Y'_{lac,glc}$, $Y'_{NH3,glc}$ $Y'_{O2,glc}$ and other yields provide a good indication of a stable culture environment. Under these conditions, Equation may be used with confidence to estimate changes in the viable cell density.

Product Formation

Specific product formation rate

The characteristic parameter for product formation is the specific product formation rate, q_P. Equations and the methods of analysis introduced for metabolic byproducts also apply to secreted protein products.

Constant q_P in batch culture

The general formula for q_P for batch growth in a stirred vessel is given by Equation, where P is the extracellular product concentration. Frequently, q_P is relatively constant during batch growth. One example is monoclonal antibody production by hybridoma cells. If it is assumed that q_P does not vary with time, then Equation can be integrated to yield Equation:

$$P = q_P \int_0^t n_v \, dt$$

The integral represents the area under the viable cell curve (n_v versus time). If q_P is indeed constant during batch growth, then a plot of product concentration versus the area under the viable cell curve will yield a straight line with slope equal to q_P. This provides an easy approach for determining q_P. The integral may be evaluated by fitting the viable cells versus time curve with a polynomial and performing the integration analytically. Alternatively, the integral may be approximated by a sum of rectangles or trapezoids.

Systems with product retention

Reactors may be operated with product retention as well as cell retention. This allows for product recovery at high concentration. Retention may be accomplished by using a membrane with a pore size that is too small to allow passage of product but large enough to allow passage of metabolites and waste products. Capsules that retain cells can also retain product. In any type of product-retention system, retention may not be 100%. This does not complicate analysis as long as the etention fraction is known. For a constant-volume reactor with product retained via a membrane, a mass balance on product gives:

$$q_P = \frac{D(\alpha_P P - P_F) + (dP/dt)}{n_v}$$

where P is the concentration of product retained in the reactor and α_P is the ratio of the product concentration in the outlet stream to the product concentration in the reactor ($1 - \alpha_P$ is the retention fraction). A drop in q_P at high P levels may indicate feedback inhibition by product.

Product yield on medium components

We can define an apparent yield of product from substrate ($Y'_{P,S}$) as a measure of the efficiency of product formation that can be used to compare cells grown in different culture systems or in different media. Product P is a cellular product such as monoclonal antibody, while substrate S may be serum, glucose, glutamine, oxygen or any other important substrate. Analysis of the overall yield (from endpoint calculations) may allow for comparison of production efficiency between different reactors and culture conditions. Analysis of the yield at various time points allows for detection of changes in the mechanisms of product formation. Product yields are useful in identifying product degradation in culture (as evidenced by a decrease in $Y'_{P,S}$ for all substrates).

Immobilized-cell reactors

It is also relevant to consider the product yield on substrate in immobilized-cell reactors. The apparent yield of product from substrate, $Y'_{P,S}$, can be monitored over time to detect changes in the yield. If other yields are relatively constant, then changes in $Y'_{P,S}$ may indicate changes in the specific production rate (q_P). However, if some or all of the other calculated metabolic yields change, then it would be very difficult to attribute changes in production rate directly to changes in q_P. Inhibition of substrate consumption or growth by waste products may contribute to a decrease in product yield.

Concluding Remarks

The growth and metabolic parameters described here (see Table 8.3) may be used directly to compare responses by different cells in a given system or by the same cells in different culture environments. They can also be incorporated into metabolic models to provide a better understanding of the processes governing cell growth, metabolism and product formation.

Table 8.3 Glossary of Terms

Term	Full extension
C_{LDH}	Concentration of lactate dehydrogenase (LDH) in culture (IU 1-l)
Co_2	Dissolved oxygen concentration (mM)
Co_2 F	Dissolved oxygen concentration in reactor feed stream (mM)
D	Reactor dilution rate (= F/V) (h^{-1})
F	Constant-volume reactor volumetric flow rate ($F_i = F_0 = F$) (l h^{-1})
f_{Fv}	Viable fraction of free cells in microcarrier culture (dimensionless)
F_i	Reactor inlet medium volumetric flow rate (l h^{-1})
F_o	Reactor outlet medium volumetric flow rate (l h^{-1})
f_v	Fraction of viable cells (dimensionless)
H	Henry's law constant for oxygen (mmHg mM^{-1})
k_A	Specific death rate (h^{-1})
k_j	Specific dead cell lysis rate (h^{-1})
k_{La}	Volumetric oxygen mass transfer coefficient (h^{-1})
k_R	Rate of cell release from microcarriers (h^{-1})
M	Metabolite concentration in reactor (mM)
m_{ATP}	ATP required for cell maintenance (mmol $cell^{-1}$ h^{-1})
M_F	Metabolite concentration in reactor feed stream (mM)
M_o	Metabolite concentration in reactor outlet stream (mM)
n	Total cell concentration (cells ml^{-1})
n_A	Concentration of attached cells in microcarrier culture (cells ml^{-1})
n_d	Dead cell concentration (cells ml^{-1})
n_e	Effective total cell concentration if no lysis occurred (cells ml^{-1})

(Continued)

TABLE 8.3 GLOSSARY OF TERMS (*CONTINUED*)

Term	Full extension
n_F	Concentration of free (unattached) cells in microcarrier culture (cells ml^{-1})
n_0	Total cell concentration in cell-retention reactor outlet stream (cells mi^{-1})
n_v	Viable cell concentration (cells ml^{-1})
n_{Vave}	Average viable cell concentration in the interval between samples (cells ml^{-1})
P	Product concentration in reactor (mM)
P_o	Oxyen partial pressure in the gas phasc (mmIIg)
P_{O2}	Molecules of ATP generated per molecule oxygen consumed (dimensionless)
P_F	Product concentration in reactor feed stream (mM)
q_{Alv}	Specific ATP production rate (mmol $cell^{-1}$ h^{-1})
Q_M	Volumetric production rate of M ($mmol^{-1}$ h^{-1})
q_u	Specific formation rate of M (mmol $cell^{-1}$ h^{-1})
q_o	Specific oxygen consumption rate (mmol $cell^{-1}$ h^{-1})
q_Y	Specific production rate of P (mmol $cell^{-1}$ h^{-1})
q_s	Specific consumption rate of S (mmol $cell^{-1}$ h^{-1})
S	Substrate concentration in reactor (mM)
S_F	Substrate concentration in reactor feed stream (mM)
t	Culture time (h)
V	Reactor volume (l)
Y'_{nATP}	Apparent yield of cells from ATP (cells $mmol^{-1}$)
F_{jS}	Apparent yield of cells from substrate S (cells $mmol^{-1}$)
F_{pS}	Apparent yield of product P from substrate S (mrnol $mmol^{-1}$)
Y_{nATP}	True yield of cells from A'I'P (cells $mmol^{-1}$)
α_p	Ratio of the product concentration in the outlet stream to the product concen tration in the reactor for a system with product retention (dimensionless)
α_s	Ratio of the cell concentration in the outlet stream to the cell concentration inthe reactor for a system with cell retention (= F_2/F_o in Figure 4.2.3) (dimension less)
α_{sd}	Ratio of the dead cell concentration in the outlet stream to the dead cell concentration in the reactor (dimensionless)
α_{sv}	Ratio of the viable-cell concentration in the outlet stream to the viable-cell concentration in the reactor (dimensionless)
β_d	Fraction of yLDH retained by dead cells (dimensionless)
γ_{LDH}	LDHcontent per cell (IU $cell^{-1}$)
μ	True specific growth rate (h^{-1})
μ_{app}	Apparent specific growth rate (h^{-1}).
μ_{app}	Viable cell-derived apparent specific growth rate (h^{-1})
μ_A	True specific growth rate of attached cells in microcarrier culture (h^{-1})
μ^A_{App}	Apparent specific growth rate of attached cells in microcarrier culture (h^{-1})
μ^e_{App}	Apparent specific growth rate based on effective total cell concentration (h^{-1})
μ_F	True specific growth rate of free cells in microcarrier culture (h^{-1})

MODELLING

A kinetic model of mammalian cells is a quantitative description of the main phenomena that have an influence on the growth, death and metabolic activities of cells. In its simplest form a model consists of a set of mathematical relationships between the different cellular rates

– growth, death, nutrient uptake and metabolite production – and the composition of the culture medium. When transferred to a computer, it provides a simulation of the time variation of the different components of the culture medium. More elaborate models, potentially capable of identifying limiting metabolic steps during biosynthesis of the desired product, may also be designed to represent changes in the intracellular content, the metabolic pathways or the cell physiology as a function of culture conditions.

As one of its major interests, a model represents an efficient tool for the kinetic analysis of cellular processes. It is able to account for the main phenomena that may simultaneously control the activities of cells. As such, depending on the culture conditions, composition of the medium and whether there is batch or continuous mode of operation, it can be used first to identify the rate-limiting factors and then to characterize quantitatively their relative importance. For instance, with a model it is possible to evaluate the kinetic effect of a depletion of glucose, glutamine and other amino acids or of an accumulation of ammonia and lactate on the rates of cell growth and death.

Because of their predictive capabilities, models are also essential tools in modern biochemical engineering for the design of processes and the optimization of media and reactor operational parameters in batch or continuous operation. They can also serve in the development of software sensors to estimate on-line the time variation of the medium composition.

The construction of a kinetic model for an animal cell culture involves several steps: a kinetic analysis of the experimental results with the formulation of hypotheses on the nature of the rate-limiting steps; the choice of rate expressions describing the influence of these phenomena on the cellular processes; evaluation of parameter values; and validation of the model with different experimental results. In this section a general methodology is described for the modelling of cell cultures, and the procedure is illustrated on the kinetics of a hybridoma cell. (For a summary of terms used).

Background for the Modelling of Mammalian Cell Cultures

A kinetic model consists of a set of mathematical expressions that relate the rates of cellular growth and metabolism to the composition of the medium. With mammalian cells, the rates usually measured are the rates of cell growth and death, the rates of uptake of the main nutrients, glucose and glutamine, the rates of production of the main metabolites, lactate and ammonia, and the rate of secretion of proteins. Most of the medium components, depending on their concentration, may have an influence on the metabolic activities of cells. Among the most frequently observed rate-limiting effects are the depletion of glucose and glutamine and the accumulation of ammonia and lactate.

An example of a kinetic model that takes into account the effect of these four components - glucose, glutamine, lactate and ammonia - is presented below. The different rate expressions it contains have been found correctly to simulate batch or continuous cultures of several mammalian cell lines.

Specific rate of cell growth:

$$\mu = \mu_{max}\left(\frac{[\mathrm{Glc}]}{K_{\mathrm{Glc}}+[\mathrm{Glc}]}\right)\left(\frac{[\mathrm{Gln}]}{K_{\mathrm{Gln}}+[\mathrm{Gln}]}\right)\left(\frac{1}{1+([\mathrm{Lac}]/K_{\mathrm{Lac}})}\right)\left(\frac{1}{1+([\mathrm{NH_4}]/K_{\mathrm{NH_4}})}\right)$$

Specific rate of cell death:

$$k_{\mathrm{d}} = A_{\mathrm{d}}\left\langle\left(\frac{1}{1+([\mathrm{Gln}]/C_1)}\right)+\left(\frac{1}{1+([\mathrm{Glc}]/C_2)}\right)+k_1[\mathrm{NH_4}]+k_2[\mathrm{Lac}]\right\rangle$$

Specific rate of glucose consumption:

$$\nu_{\mathrm{Glc}} = Y_{\mathrm{Glc/X}}\cdot\mu + m_{\mathrm{Glc}}$$

Specific rate of lactate production:

$$\pi_{\mathrm{Lac}} = Y_{\mathrm{Lac/X}}\cdot\mu + m_{\mathrm{Lac}}$$

Specific rate of glutamine consumption:

$$\nu_{\mathrm{Gln}} = Y_{\mathrm{Gln/X}}\cdot\mu + m_{\mathrm{Gln}}$$

Specific rate of ammonia production:

$$\pi_{NH4} = Y_{NH4/X} \cdot \mu + m_{NH4}$$

Specific rate of antibody production:

$$\pi_{MAbs} = Y_{MAbs/X} \cdot \mu + m_{MAbs}$$

Rate of glutamine degradation:

$$r_{Gln} = k_{deg}[G\ln]$$

Specific rate of cell growth:

The specific rate of cellular growth, μ, is defined as the number of new cells produced per unit (e.g. billion) of living cells present in the culture medium per unit time (e.g. hour).

For a given medium a cell line can be characterized by a maximum specific growth rate, μ_{max}, which is the observed growth rate in the absence of any limitations by nutrients or any inhibition by metabolites. This maximal growth rate is related to the doubling time (t_d) of the cell by the relationship:

$$\pi_{MAbs} = Y_{MAbs/X} \cdot \mu + m_{MAbs}$$

During culture the specific growth rate usually decreases because of either depletion of essential nutrients or accumulation of inhibitory metabolites. Equation represents such a variation of specific growth rate as a function of the concentration of glucose, glutamine, lactate and ammonia in the medium. In this kinetic law the maximum specific growth rate is multiplied by four terms that describe the rate-limiting effect of each of the components.

The parameters introduced – K_{Glc}, K_{Gln}, K_{Lac}, K_{NH4} - give the range of concentrations where either the nutrient becomes limiting or the metabolite becomes inhibitory. By modulating the values of these parameters the model can account for differences in cell sensitivities towards nutrient depletion and product inhibition.

Specific rate of cell death

The specific rate of cellular death, k_d, is defined as the number of dying cells per unit (e.g. billion) of living cells present in the culture medium per unit time (e.g. hour).

As a first approximation k_d has often been considered as a constant. Yet more detailed kinetic analyses have shown that the specific rate of cell death is also affected by the chemical composition of the medium and several physicochemical parameters, such as pH, temperature and osmotic pressure. It is often lowest at the start of the culture, and then gradually increases due either to depletion of essential nutrients or accumulation of inhibitory metabolites.

A rate expression for cell death is given in Equation. With its four terms it expresses the possible increase in the cell death rate due to limitations in glucose and glutamine or accumulations in lactate and ammonia. In this case, the different contributions are additive in order to take into account the effect of each component: if the expression relative to one substrate becomes equal to zero, the effects of the other components remain visible in the calculation of the specific death rate. By proper adjustment of the values of the four parameters C_1 C_2, κ_1 and κ_1 it is possible to account for differences in death kinetics from one cell line to another.

Specific rate of nutrient uptake

The specific rate of nutrient uptake, *v*, is defined as the number of millimoles of nutrient consumed per unit (e.g. billion) of living cells present in the culture medium and per unit time (e.g. hour).

For most cell lines it has been found to increase linearly with the specific growth rate. Thus the specific rate of glucose uptake is often expressed as a function of μ by Equation, which contains two parameters: the non-growth-associated specific glucose consumption rate, m_{Glc} (mmol/glucose 10^{-9} cells h^{-1}), and the glucose to biomass conversion yield, $Y_{Glc/X}$ (mmol glucose 10^{-9} cells). A similar expression is applicable to the specific rate of glutamine uptake.

Specific rate of metabolite and protein production

The specific rate of metabolite or protein production, π, is defined as the number of millimoles or milligrams of product excreted per unit (e.g. billion) of living cells present in the culture medium and per unit time (e.g. hour).

For most cell lines it is found to increase with the specific growth rate. Thus the specific rates of lactate production can be expressed as a function of *x*, which contains two parameters: the non-growth-associated specific lactate production rate, m_{Lac} (mmol lactate 10^{-9} cells hr^{-1}), and the lactate to biomass stoichiometric yield, $Y_{mac/X}$ (mmol lactate 10^{-9} cells). A similar expression is often applicable to the specific rate of ammonia production and antibody secretion.

Rate of glutamine decomposition

The spontaneous decomposition in the medium of glutamine into ammonia can be represented by a first-order rate process with respect to the glutamine concentration.

Method for Kinetic Model Construction

Experimental investigations

As one objective of model construction is to obtain the best fit between model simulations and experimental results, appropriate kinetic data have to be obtained. Experiments can be performed in different systems:

- *Batch cultures* are the simplest to perform, either in shake flasks or small bio-reactors. However, as the concentration of all the nutrients and metabolites changes simultaneously with time, it is relatively difficult to assess the precise influence of a single medium component on cell kinetics
- *Continuous cultures* require more sophisticated equipment and long-term operation extending to several weeks. But, as a major advantage, nutrient and metabolite concentrations can be maintained constant for several days, which allows a more precise analysis of the influence of medium composition on cellular activity

When investigating cell kinetics, it is very important to control precisely the physicochemical parameters of the medium, such as temperature, pH, dissolved oxygen and osmotic pressure. It is also important to define precisely the state of the inoculum, which may have a significant effect on the progress of the culture. During the culture, one measures at regular time intervals the concentrations of living and dead cells, of the major nutrients and metabolites and of excreted proteins.

Example

The procedure of kinetic data analysis and model construction is illustrated for a hybridoma culture (cell line VO 208) in a batch system. The medium used was RPMI 1640 + 5% (v/v) foetal calf serum (FCS) + 2% (v/v) minimum essential medium (MEM) amino acids + 1% non-essential amino acids and initial glucose and glutamine concentrations of 13 mM and 4.5 mM, respectively.

The batch culture was carried out in a bioreactor with a working volume of 1 l. Forty-eight hours after the last Roux bottle inoculation, the cells were inoculated at about 2×10^8 viable cells l^{-1} in the bioreactor.

The pH was maintained at a value of 7 with 0.2 M NaOH solution and gaseous CO_2. The temperature was set at 37°C and the oxygen supply regulated at 50% of air saturation with gaseous air and nitrogen. The rotating speed was 50 rpm.

During the culture, the concentrations of living and dead cells (by the Trypan blue exclusion method) were measured using a haemocytometer: glucose, lactate and glutamine by enzymic methods; ammonia with a selective electrode; and monoclonal antibodies by ELISA assay.

The time variations of these medium compounds for the given example are presented.

Kinetic analysis

Prior to the construction of the kinetic model, one has to perform a detailed analysis of the experimental data in order to identify the main rate-limiting effects and the relationships that may exist between the different kinetic variables. A procedure for data analysis is described and illustrated for the previously obtained experimental results.

Identify the rate-limiting nutrients

The time variation of the different concentrations in the culture medium shows the classical mammalian cell kinetics of a batch culture: a growth phase with a maximal cell concentration of 8×10^8 viable cells l^{-1}, followed, after 100 h of culture, by a death phase. Rapid death occurs when glutamine is completely consumed in the medium, which indicates that glutamine limitation is responsible for the cessation of cell

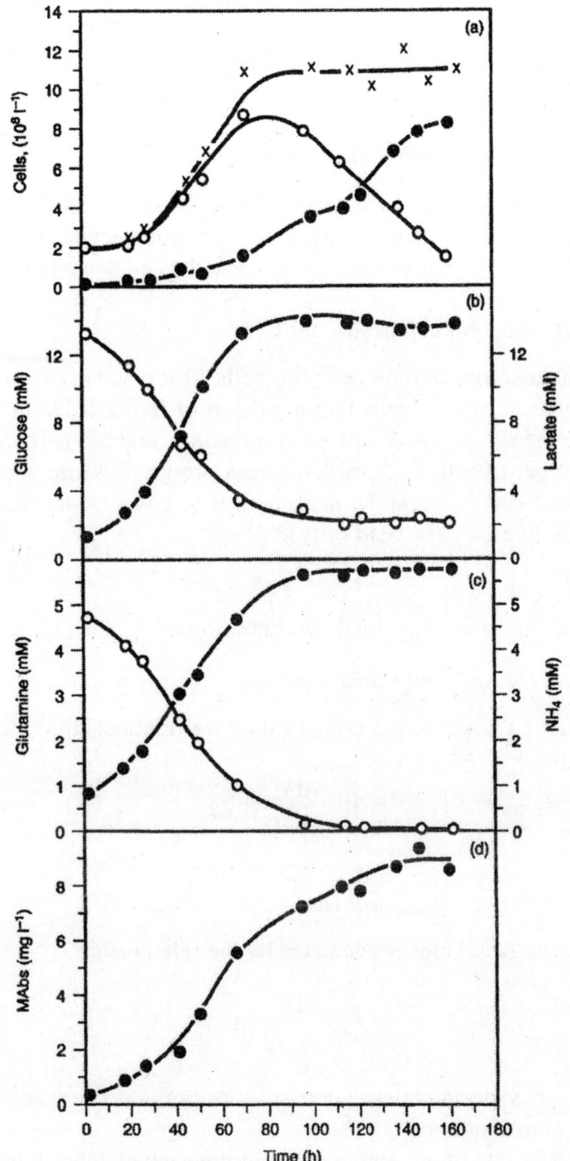

Fig. 8.9 Evolution with time of total ((a), x), viable ((a), ○) and Trypan blue dead cells ((a), ●) and glucose ((b), ○), lactate ((b), ●), glutamine ((c), ○), ammonia ((c), ●) and monoclonal antibody ((d), ●) concentrations during VO 208 hybridoma batch culture.

growth and increase in cell death rate. The glucose level progressively decreases and reaches a residual value near 2 mM. At this level, the rate-limiting effect of glucose on cell growth and death is probably not significant.

Identify the rate-limiting metabolites

The lactate concentration increases up to 14 mM, and thc ammonia concentration up to 5.5 mM. At these levels, both lactate and ammonia can inhibit the cell growth rate or increase the cell death rate.

Calculate the total cell production

The measured concentration of living cells is the result of two processes: the formation of new cells and the death of living cells. In order to evaluate the actual rate of new cell production, one has to calculate the total number of cells (X_t) that have been produced during the culture. As long as cell lysis can be neglected, it is given by the sum of the measured live (X_v) and dead cells (X_d):

$$X_t = X_v + X_d$$

The time variation of the total concentration of cells is plotted.

Calculate the specific growth rate

The specific growth rate is calculated by the relationship:

$$\mu = \frac{dX_t}{X_v dt} (h^{-1})$$

Calculate the specific death rate

The specific death rate is calculated by the relationship:

$$k_d = \frac{dX_m}{X_v dt} (h^{-1})$$

Calculate the specific rates of nutrient consumption The specific rate of glucose consumption:
The specific rate of glutamine consumption (which takes into account glutamine decomposition):

$$\nu_{Glc} = \frac{-d[Glc]}{X_v dt} (\text{mmol } 10^{-9} \text{cell h}^{-1})$$

Calculate the specific rate of metabolites and antibody production

$$\nu_{Gln} = \frac{-d[Gln]}{X_v dt} - \frac{k_{deg}[Gln]}{X_v} (\text{mmol } 10^{-9} \text{cell h}^{-1})$$

The specific rate of lactate production:

$$\pi_{Lac} = \frac{d[Lac]}{X_v dt} (\text{mmol } 10^{-9} \text{cell h}^{-1})$$

The specific rate of ammonia production:

$$\pi_{NH_4} = \frac{d[NH_4]}{X_v dt} - \frac{k_{deg}[Gln]}{X_v} (\text{mmol } 10^{-9} \text{cell h}^{-1})$$

The specific rate of antibody production:

$$\pi_{MAbs} = \frac{d[MAbs]}{X_v dt} (\text{mmol } 10^{-9} \text{cell h}^{-1})$$

Plot the different specific rates as a function of time and analyse the evolution of the metabolic activities of cells during the culture
The resulting curves clearly indicate the existence of two culture periods:

- **Phase 1** lasts about 30 h and is where the different rates increase. This corre sponds to the classically observed lag phase, where cells progressively adapt to their new environment.
- **Phase 2** is from 40 to 1.40 h, where all the specific rates progressively decrease except the cellular death rate, which continues to rise. This reduction in the specific rates of growth and metabolism is mainly due to glutamine limitation. However, the accumulation of lactate and ammonia may also have a kinetic effect.

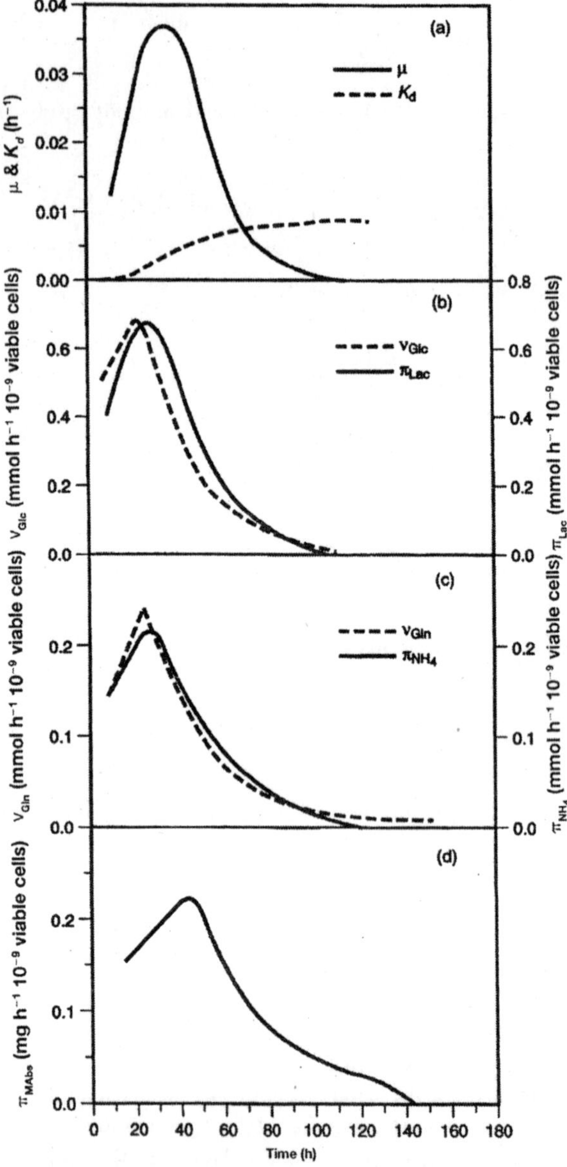

Fig. 8.10 Evolution with time of the specific growth and death rates (a), the specific consumption rates of glucose (b) and glutamine (c) and the specific production rates of lactate (b), ammonia (c) and monoclonal antibodies (d) in VO 208 batch culture.

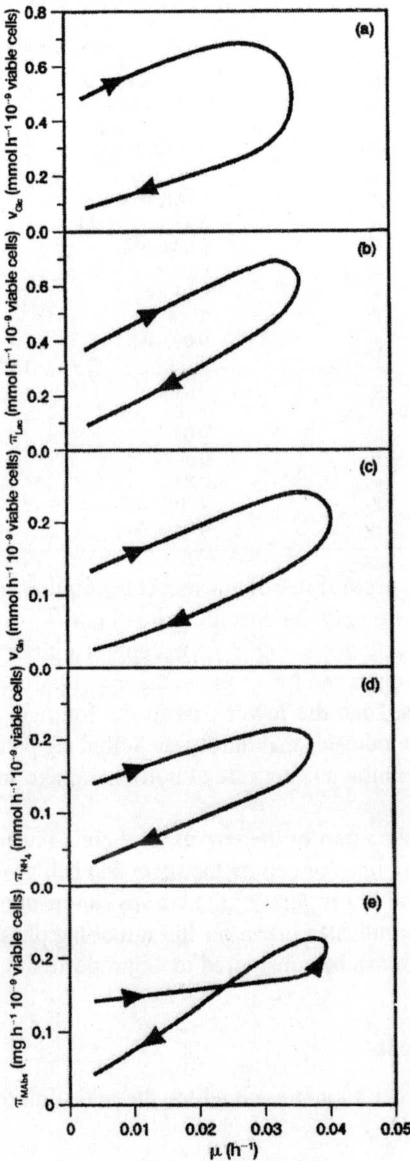

Fig. 8.11 Relationships between specific glucose (a) and glutamine (c) consumption rates, lactate (b), ammonia (d) and monoclonal antibody (e) production rates and specific growth rate during lag and growth phases of VO 208 batch culture.

TABLE 8.4 RANGE OF PARARNETER VALUESA

	Parameter	Order of magnitude of parameter values
Growth	μ_{max}	0.02–0.125 h^{-1}
	K_{Glc}	0.1–1 mM
	T_{Gln}	0.06–0.3 mM
	K_{Lac}	8–12 000 mM
	K_{NH_4}	1–45 mM
Death	A_d	0.008–0.08 h^{-1}
	C_1	5 x 1w–0.1 mM
	C_2	0.004 mM
	κ_1	0.03–0.05 mM^{-1}
	$\kappa 2$	0.01 mM^{-1}
Glucose consumption	$Y_{Glc/x}$	2–16
	m_{Glc}	0.007–0.5
Lactate production	$Y_{Lac/x}$	4–15
	m_{Lac}	0.01
Glutamine consumption	$Y_{Gln/X}$	1–3
	m_{Gln}	0.04
Ammonia production	$Y_{NH_4/X}$	0.5–2
	m_{NH_4}	0.03
Antibody production	$Y_{MAbs/X}$	4–10
	m_{MAbs}	0.04–0.06

Represent the specific rates of nutrient consumption and metabolite production as a function of the specific growth rate

All the curves obtained in Fig. 8.12, except for antibody production, show two distinct upper and lower parts: the upper part corresponds to the initial lag phase, and the lower part to the following growth and death periods. This indicates that during the initial lag phase, at a given growth rate, cells exhibit a faster rate of nutrient uptake and metabolite production.

Analyse the lower part of the curves. Is there a linear relationship between the specific rates of cell metabolism and cell growth?

In Fig. 8.12, the lower parts of the curves can be approximated to straight lines. This indicates that after the initial lag phase, the rate of cellular metabolism can be considered to be proportional to the rate of cellular growth.

Model development

We will only consider a kinetic model for the main culture phase after the initial lag period.

Select a rate expression for cellular growth

As suggested from the previous kinetic analysis, glutamine, lactate and ammonia have the main limiting effects on cellular growth. Thus, according to the model database, a kinetic expression with the three rate-limiting effects is selected:

$$\mu = \mu_{max}\left(\frac{[\text{Gln}]}{K_{\text{Gln}}+[\text{Gln}]}\right)\left(\frac{1}{1+([\text{Lac}]/K_{\text{Lac}})}\right)\left(\frac{1}{1+([\text{NH}_4]/K_{\text{NH}_4})}\right)$$

Select a rate expression for cellular death

Glutamine depletion as well as lactate and ammonia accumulation are also responsible for the observed increase in the specific rate of cell death. Again a three-term rate expression is selected:

$$k_d = A_d\left\{\left(\frac{1}{1+([\text{Gln}]/C_1)}\right)+\left(\frac{1}{1+([\text{Glc}]/C_2)}\right)+k_1[\text{NH}_4]+k_2[\text{Lac}]\right\}$$

Select the rate expressions for nutrient uptake

As linear relationships have been found for both glucose and glutamine uptake rate as a function of the specific growth rate, the kinetic expressions proposed in Equations can be considered applicable:

$$\nu_{\text{Glc}} = Y_{\text{Glc/X}}\cdot\mu + m_{\text{Glc}}$$

$$\nu_{\text{Glc}} = Y_{\text{Gln/X}}\cdot\mu + m_{\text{Gln}}$$

Select the rate expressions for metabolite and antibody production

Linear relationships are also applicable to the specific rates of lactate, ammonia and antibody production:

$$\pi_{\text{Lac}} = Y_{\text{Lac/X}_v}\cdot\mu + m_{\text{Lac}}$$

$$\pi_{\text{NH}_4} = Y_{\text{NH}_4}/_{\text{NH4/X}_v}\cdot\mu + m_{\text{NH}_4}$$

$$\pi_{\text{MAbs}} = Y_{\text{MAbs/X}_v}\cdot\mu + m_{\text{MAbs}}$$

Write the mass balance equations for the different species.

In order to simulate the time-course of a batch culture with the previously introduced rate expressions, one writes the mass balance equation for each of the considered species. They express the fact that in a batch culture the rate of variation in the concentration of a given species in the culture medium is equal to the rate of production or disappearance of the species. One thus obtains the following set of differential equations:

$$\frac{dX_v}{dt} = (\mu - k_d)X_v$$

$$\frac{d[\mathrm{Glc}]}{dt} = -\nu_{\mathrm{Glc}} X_v$$

$$\frac{d[\mathrm{Lac}]}{dt} = \pi_{\mathrm{Lac}} X_v$$

$$\frac{d[\mathrm{Gln}]}{dt} = -k_{\mathrm{deg}}[\mathrm{G\,ln}] - \nu_{\mathrm{Gln}} X_v$$

$$\frac{d[\mathrm{NH_4}]}{dt} = -k_{\mathrm{deg}}[\mathrm{G\,ln}] - \pi_{\mathrm{NH_4}} X_v$$

$$\frac{d[\mathrm{MAbs}]}{dt} = \grave{A}_{\mathrm{MAbs}} X_v$$

Integrate Equations by a numerical method

Several standard procedures are available numerically to integrate these equations and thus calculate the time variations of the different concentrations. The Runge-Kutta method is generally advised.

Parameter identification

Finally one has to determine the values of the parameters introduced in the different rate expressions. The objective is to obtain the best agreement between the calculated time variations of the medium composition and the measured experimental results. There are two ways to evaluate the parameters: some can be calculated directly from the experimental results and others are evaluated by a curve-fitting procedure.

Evaluate the values of the parameters in the rate expressions of the specific rates of nutrient uptake and metabolite or protein productions.

Each of the five linear relationships between the specific rate of cell metabolism and the specific rate of cell growth contains two parameters: a stoichiometric coefficient Y and a non-growth-associated specific rate m. From the plots shown one can thus evaluate the values of ten of the parameters as the slope of the straight lines and their intercept with the y-axis. The determined values are reported.

Evaluate the values of the other parameters by a curve-fitting procedure.

The eight parameters that have been introduced in the rate expressions for cellular growth and death remain to be determined. As a first estimation one can take the lower limit of the usual range of values found for other cell lines as indicated.

With these first values we do not obtain a correct simulation of the experimental results. Thus the values of the parameters are progressively improved by trial and error. One modifies the value of one or several of the parameters and determines if these results are an improvement in the fit between the model and the experimental results. The procedure is repeated until a satisfactory agreement is achieved. The resulting values are reported in Table and the results of the final simulation are presented.

By this iterative procedure one observes that model predictions are very sensitive to the values of some of the parameters, which

TABLE 8.5 PARAMETER VALUES FOR THE VO 208 HYBRIDOMA CELL LINE

	Parameter	Parameter values
Growth	μ_{max}	0.058 h^{-1}
	K_{Gln}	0.1 mM
	K_{Lac}	55 mM
	K_{NH_4}	10 mM
Death	A_d	0.008 h^{-1}
	C_1	0.1 mM
	κ_1	0.05 mM^{-1}
	κ_2	0.014 mM^{-1}
Glucose consumption	$Y_{Glc/x}$	9
	m_{Glc}	0.018
Lactate production	$Y_{Lac/x}$	13
	m_{Lac}	0.01
Glutamine consumption	$Y_{Gln/X}$	2.5
	m_{Gln}	0.037
Ammonia production	$Y_{NH_4/X}$	1.7
	m_{NH_4}	0.03
Antibody production	$Y_{MAbs/X}$	4.5
	m_{MAbs}	0-06

as a consequence have to be determined with great precision. On the contrary, for others, approximate values are sufficient because model predictions are not significantly affected by variations in these parameter values.

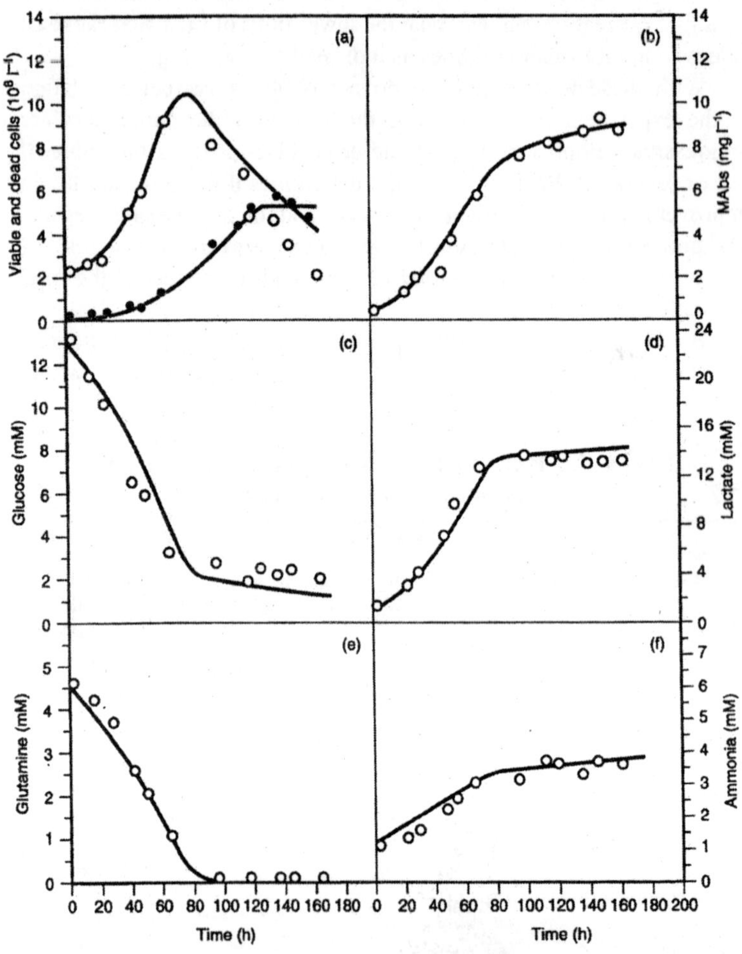

Fig. 8.12 Experimental (○, ●) and theoretical (—) evolutions with time of viable. ((a), ○) and dead ((a), ●) cell density and monoclonal antibody (b), glucose (c), lactate (d), glutamine (e) and ammonia (f) concentrations during VO 208 batch culture (RPM1 containing 12.5 mM glucose, 4,5 mM Gln and 5% (v/v) FCS).

USE OF THE MODEL FOR THE EVALUATION OF MATE-LIMITING FACTORS

The model proposed for mammalian cell cultures provides a description of the possible influence of four of the main medium components – glucose, glutamine, lactate and ammonia – on the rates of cellular growth, death and metabolism. It contains kinetic terms that quantify the influence of each of the components either in reducing the rate of cellular growth or in increasing the rate of cell death.

During a batch or continuous culture, where the concentrations of all four species change with time, the observed reduction in growth rate or increase in death rate is generally the result of the simultaneous influence of several of these effects. From a simple kinetic analysis it is seldom possible to evaluate the relative effects of substrate depletion or metabolite accumulation on the observed kinetics. With the kinetic model as described, it becomes straightforward to characterize quantitatively the kinetic effects of the different medium components, as outlined here.

The procedure is based on the previous batch hybridoma culture, for which three growth-limiting effects - by glutamine, lactate and ammonia - were considered in the model.

Calculate the growth rate reduction factors by nutrients and metabolites

The kinetic expression for the specific rate of cellular growth Equation contains three terms that each express a possible reduction of the specific growth. rate with respect to μ_{max}. These growth-rate-reducing factors are evaluated as:

$$[\text{Gln}]/(K_{\text{Gln}} + [\text{Gln}])\text{for glutamine}$$

$$1/(1+[\text{Lac}]/K_{\text{Lac}})\text{for lactate}$$

$$1/(1+[\text{NH}_4]/K_{\text{NH}_4})\text{for ammonia}$$

From the measured (or the model-calculated) time variations of the concentrations of the three different species, one estimates the values of the corresponding factors: they are equal to unity when the considered substrate or product is non-limiting and they decrease when the substrate level becomes too low or the metabolite level too high.

Calculate the factors increasing the death rate by nutrients and metabolites

The kinetic expression for the specific rate of cellular death contains three terms that each express a possible increase of the specific death rate and can be evaluated as follows:

$$1/(1+[\mathrm{Gln}]/C_1)\text{for glutamine}$$
$$\kappa_2 \cdot [\mathrm{Lac}])\text{for lactate}$$
$$\kappa_1 \cdot [\mathrm{NH_4}] \text{ for ammonia}$$

From the measured or simulated time variations of the concentrations of the three different species, one estimates the values of the three different factors: they are equal to zero when the considered substrate or product is non-limiting and they increase when the substrate level becomes too low or the byproduct concentration too high.

Compare the kinetic effects of the nutrients and metabolites

The time variations of the different factors have been calculated and are represented in Figs., together with the variation of the specific rates of cell growth and death. These variations clearly indicate that during the first 70 h the slight decrease in growth rate and increase in death rate can be essentially attributed to the action of ammonia and lactate. After 80 h the sharp decrease in growth rate and increase in death rate result from the simultaneous effect of ammonia, lactate and glutamine, but glutamine depletion has the most dominant effect.

From a single batch culture it is difficult to identify precisely the kinetic effects of the four considered medium components. It would be necessary to study the kinetics of several cultures run at different initial medium concentrations of glucose and glutamine. A better quantification of the respective effects of lactate and ammonia on cell growth and death would also require additional studies on culture kinetics when adding different levels of lactate or ammonia to the initial culture medium. An alternative method would be to grow the cells in continuous mode and change separately the concentrations of the medium components in the feeding medium.

The extent of cell death is determined by measuring the cells unable to exclude the Trypan blue dye. In prolonged continuous

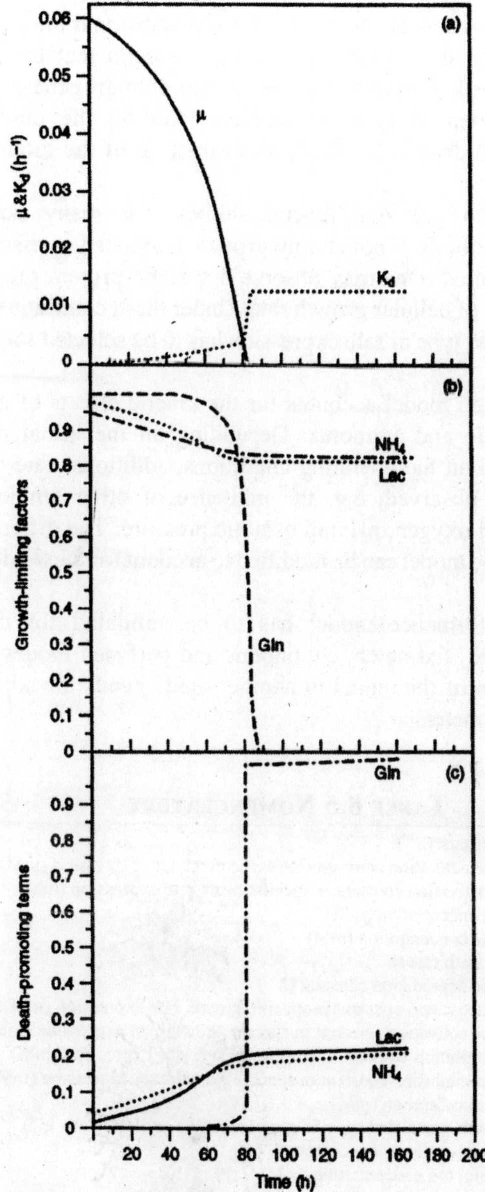

Fig. 8.13 Modelling evolution with time of the specific growth and death rates (a) and their limiting (b) and promoting (c) factors during a batch culture.

culture or in unfavourable environments the additional phenomenon of cell lysis may also occur. The number of cells that have lysed can be determined by measuring the lactate dehydrogenase (LDH) concentration released into the medium. Adding the number of visible and lysed dead cells yields an evaluation of the total rate of cell death.

According to previous kinetic studies, for many cells the production of antibody is not simply growth associated as assumed in the model presented. One may observe a specific protein production rate independent of cellular growth rate. Under these conditions a non-growth-associated type of rate expression has to be selected for protein production.

The suggested model accounts for the kinetic effects of glucose, glutamine, lactate and ammonia. Depending on the initial medium composition and on the operating conditions, additional rate-limiting effects may be observed, e.g. the influence of other amino acids, serum, dissolved oxygen, pH and osmotic pressure. The different rate expressions in the model can be modified to account for these additional effects.

A high-performance model has to be validated for different procedures: batch, fed-batch, continuous and perfused modes. Often, the extrapolation of the model to another mode needs the adjustment of different parameters.

TABLE 8.5 NOMENCLATURE

A_d	Death constant (h^{-1})
C_1	Glutamine activation constant in specific death rate expression (mM)
C_2	Glucose activation constant in specific death rate expression (mM)
[Glc]	Glucose concentration (mM)
[Gin]	Glutamine concentration (mM)
k_d	Specific death rate (h^{-1})
k_{deg}	Glutamine degradation constant (h^{-1})
K_{Glc}	Glucose activation constant in specific growth rate expression (mM)
K_{Gln}	Glutamine activation constant in specific growth rate expression (mM)
K_{Lac}	Lactate inhibition constant in specific growth rate expression (mM)
K_{NH_4}	Ammonia inhibition constant in specific growth rate expression (mM)
[Lac]	Lactate concentration (mM)
m	Non-growth-associated specific rates (mmol or mg 10^{-9} cells h^{-1})
[MAbs]	Monoclonal antibody concentration (mg l^{-1})
$(NH_4^+]$	Ammonium ion concentration (mM)
r_{Gln}	Rate of glutamine degredation (mmol l^{-1} h^{-1})
X_v	Viable cell concentration (10^9 cells l^{-1})
X_m	Dead cell concentration (10^9 cells l^{-1})
Y	Growth-associated yield (mrnol or mg 10^{-9} cells)

TABLE 8.5 NOMENCLATURE (*CONTINUED*)

κ_1	Lactate activation constant in specific death rate expression (mM^{-1})
κ_2	Ammonia activation constant in specific death rate expression (mM^{-1})
μ	Specific growth rate (h^{-1})
μ_{max}	Maximum specific growth rate (h^{-1})
ν_{Glc}	Specific glucose consumption rate (mmol 10^{-9} cells h^{-1})
ν_{Gln}	Specific glutamine consumption rate (mmol 10^{-9} cells h^{-1})
π_{Lac}	Specific lactate production rate (mrnol 10^{-9} cells h^{-1})
π_{NH_4}	Specific ammonia production rate (mmol 10^{-9} cells h^{-1})
π_{MAbs}	Specific antibody production rate (mmol 10^{-9} cells h^{-1})

CELL DEATH IN CULTURE SYSTEMS (KINETICS OF CELL DEATH)

Cell death in eukaryotic cells may be divided into two morphologically and biochemically distinct modes, those of apoptosis and necrosis. Necrosis is the classically recognized form of cell death that results from severe cellular insults and involves rapid cell swelling and lysis and is a process over which the cell has little or no control. Apoptosis, on the other hand, is the mode of cell death observed primarily under physiological conditions, and is an organized, preprogrammed response of the cell to changing environmental conditions. This form of cell death is characterized by cell shrinkage, nuclear and DNA fragmentation and breaking up of the cell into membrane-bounded vesicles, termed 'apoptotic bodies,' which are subsequently ingested by neighbouring cells or macrophages *in vivo.* Apoptosis is an ATP-dependent process that in some systems also requires RNA and protein synthesis. The activation of a Ca^{2+}/Mg^{2+} -dependent, zinc-inhibitable endonuclease, which cleaves the cell's DNA into fragments of 200 base pairs or multiples thereof, is the main biochemical hallmark of apoptosis.

In cell culture systems, cell death via apoptosis is quite common and can occur under a variety of circumstances. For example, growth-factor-dependent cell lines die by apoptosis following removal of the growth-promoting agent from the culture medium. Terminal differentiation of cells *in vitro* results in apoptosis, as does normal turnover of cells with a limited lifespan, e.g. neutrophils or freshly isolated thymocytes. Altered culture medium conditions, e.g. removal of zinc from the medium, or cultures that are allowed to overgrow to produce high cell densities, also induce apoptosis. Finally, cells may also be induced to undergo apoptosis by exposure to a wide range of cytotoxic agents. Under *in vivo* conditions, apoptotic cells are recognized and rapidly removed by phagocytic cells. However, in cell culture this cannot occur and instead the cells undergo secondary necrosis. Cells at this stage are Trypan blue positive. Up to this point, however, apoptotic cells maintain the ability to exclude vital dyes, and hence this method underestimates the health of a cell culture.

Reagents and solutions

Note: Lysis buffer (without proteinase K) and the TE buffer should be auto-claved to remove contaminating deoxyribonuclease activity.

- *Lysis buffer*: 10 mM EDTA, 50 mM Tris (pH 8.0) containing 0.5% AMauroyl-sarcosinc and 0.5 mg ml^{-1} proteinase K
- *TE buffer*: 10 mM Tris-HCl (pH 8.0) and 1 mM EDTA
- *Electrophoresis loading buffer*: 10 mM EDTA, 0.25% bromophenol blue and 50% glycerol
- *TBE buffer*: 2 mM EDTA (pH 8.0), 89 mM Tris and 89 mM boric acid

Procedures: Morphological Characterization of Cell Death

The two modes of cell death may be identified easily on morphological grounds. Apoptosis involves cell shrinkage, nuclear fragmentation and apoptotic bodies budding off, maintaining the ability to exclude vital dyes. Necrosis involves cell swelling, chromatin flocculation and direct cell lysis. Thus, cells undergoing necrosis rapidly lose their ability to exclude vital dyes.

Materials and equipment

- Cell fixative - a haematoxylin and eosin stain (e.g. Rapi-Diff II, Diachem Diagnostic Developments, Southport, UK)
- Trypan blue
- Cytocentrifuge

Trypan blue is a suspected carcinogen. Gloves should be worn when using all stains.

1. Cytocentrifuge cells onto glass slides. The speed of centrifugation will depend on the cell type; a typical cytocentrifuge setting would be 300 rpm for 2 min.
2. Remove glass slides and place in fixing solution for 20 s ('A' of Rapi-Diff II stain). Remove excess fluid from slide and repeat for the cytoplasmic and nuclear stains ('B' and 'C of Rapi-Diff II stain, respectively). Rinse with distilled water.

3. Examine under the light microscope.

The key morphological criteria for recognizing apoptotic cells are nuclear condensation and fragmentation. Apoptotic cells also maintain their ability to exclude vital dyes such as Trypan blue. Hence cultures with a high proportion of cells displaying nuclear condensation and fragmentation, indicative of apoptosis, should also have a relatively high number of cells with the ability to exclude vital dyes.

Necrotic cells may be recognized morphologically by cell swelling and nuclear fiocculation. Cultures with a high proportion of necrotic cells should have a relatively low number of cells with the ability to exclude vital dyes.

PROCEDURE: BIOCHEMICAL CHARACTERIZATION OF CELL DEATH

The biochemical hallmark of apoptosis is cleavage of the nuclear DNA into -200 base pair multiples. This specific DNA cleavage is thought to result from the activation of an endogenous endonuclease that cleaves at the exposed linker regions between nucleosomes. Necrosis, on the other hand, is not associated with the ordered form of DNA cleavage observed during apoptosis. Hence a biochemical examination of the DNA from cells can be used to characterize the mode of cell death occurring in cell cultures.

Isolation of DNA

- Lysis buffer
- RNase A (previously heat treated to remove contaminating deoxyribonuclease activity)
- Phenol, buffered with 0.1 M Tris-HCl (pH. 7.4)
- Chloroform/isoamyl alcohol (24:1)

Gloves should be worn throughout this procedure, both to protect the worker from phenol and chloroform and to prevent contamination of samples with deoxyribonucleases. Pipette tips and tubes should be autoclaved before use.

1. Wash cells twice in phosphate-buffered saline (PBS).
2. Resuspend cell pellets at 2×10^7 ml^{-1} in lysis buffer and incubate in a 50°C water-bath for 1 h.

3. Add RNase A to a concentration of 0.25 mg ml^{-1} and continue incubation at 50°C for 1 h.
4. Extract the crude DNA preparations twice with an equal volume of phenol (retain the aqueous layer after each extraction).
5. Extract twice with an equal volume of chloroform/isoamyl alcohol (retain the aqueous layer after each extraction).
6. Centrifuge the DNA preparations at 13 000 *g* for 15 min to separate intact from fragmented chromatin.
7. Place the supernatant (containing the fragmented chromatin) into a separate tube.
8. Precipitate the DNA in two volumes of ice-cold ethanol overnight at −70°C. The DNA may be stored in this condition for long periods.

Electrophoresis of DNA Materials and equipment

- TE buffer
- Electrophoresis loading buffer
- Agarose
- TBE buffer
- Ethidium bromide
- UV source (Transilluminator (UV: 302 nm), UVP Inc., San Gabriel, CA, USA)

Ethidium bromide is an irritant and known mutagen. Gloves should be worn throughout this procedure.

1. Recover the DNA precipitates by centrifugation at 13 000 *g* for 15 min. Allow the tube to air-dry at room temperature for 10 min and resuspend in TE buffer. Store at 4°C.
2. Add electrophoresis loading buffer in a 1:5 ratio. Place samples into a 65°C water-bath for 10 min and then maintain on ice.
3. Place samples into wells of a 1% agarose gel.
4. Electrophoresis is carried out in TBE buffer at 6 V cm^{-1} of gel.
5. After electrophoresis, soak the gel in TBE buffer containing 1 mg ml^{-1} ethidium bromide. Destain briefly in TBE buffer.

Visualize the DNA by UV fluorescence. Care should be taken not to expose skin or eyes to UV light.

The DNA isolated from apoptotic cells separates into bands of 200 or multiples of 200 base pairs, whereas DNA isolated from control cells shows relatively little degradation. The DNA isolated from necrotic cells shows no ordered DNA fragmentation.

Note: The use of DNA molecular size markers enables an estimate of the size of the fragmented DNA.

Supplementary Procedure: Purification of Apoptotic Cells

If apoptosis is only occurring at relatively low levels in cultures, it may be necessary to obtain a purified population of apoptotic cells prior to DNA isolation and electrophoresis. This may be achieved by exploiting the fact that apoptotic cells are more dense than normal cells. Hence it is possible to purify apoptotic cells by isopycnic centrifugation. Percoll can be used to create solutions of different densities. The precise Percoll densities used for isolation of apoptotic cells will depend on the cell type under investigation.

1. Prepare Percoll solutions of various densities. These densities should typically range from 1.05 to 1.08 in the case of most inanimation cells.
2. Carefully layer the Percoll solutions sequentially into a test tube. Wash cells in PBS, resuspend at a high concentration (typically 20×10^6 ml^{-1}) in PBS and place on top of the Percoll column.
3. Centrifuge at 400 *g* for 30 min in a swing-out rotor. Cells distribute to their isopycnic points, creating bands representing normal, dead and apoptotic cells.
4. Elute the different bands from the Percoll gradients. Morphological examination of these bands indicates which band is enriched for apoptotic cells. This provides an estimate for the density of the apoptotic cells.
5. Adjust the Percoll gradients, placing the solution with the estimated density of apoptotic cells at the bottom of the tube.
6. Repeat the above procedure until apoptotic cells pellet to the bottom of the test tube after centrifugation. It is possible to obtain a purified population of apoptotic cells (>90%) by this method.

Apoptosis was first described in 1972. Since then it has become apparent that this mode of cell death plays a role in a number of important physiological processes. Thus it is advantageous to be able to recognize when this mode of death is occurring. Investigations into the mechanistic aspects of apoptosis have revealed a regulatory role of different ions in this process. For example, apoptosis occurs in certain human cell lines under conditions of zinc deficiency. Likewise, apoptosis may be inhibited by the addition of zinc to the culture medium or by removal of calcium. Hence it is apparent that different culture conditions may activate or inhibit this mode of cell death. The above morphological and biochemical procedures should be used to complement each other in identifying the mode of cell death occurring in cells.

Detoxification of Cell Cultures

When the growth of a cell culture slows down at the end of the exponential growth phase, the probable reason for this Is either that the cells have consumed essential growth factors or nutrients, or that the cells are producing inhibitory or toxic metabolites. In those cases where the latter of these alternatives is predominant, it might be useful to remove the toxic metabolites in order to obtain higher cell densities or a prolonged culture lifetime. In cultures that produce a cellular

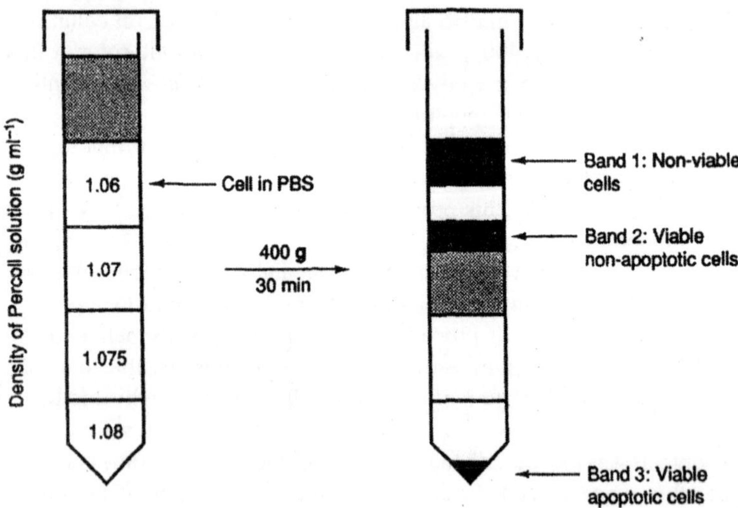

Fig. 8.14 Percoll gradient for isolation of apoptotic cells from an LH-60 cell culture.

product (e.g. monoclonal antibody or recombinant DNA product), the purpose of removing toxic metabolites might be to increase product concentration in the medium.

Procedures are given for two different methods: detoxification by dialysis and detoxification by gel filtration.

PROCEDURE: DETOXIFICATION BY DIALYSIS

Materials and equipment

- Growth medium with all supplements ***except*** serum, at least ×10 of the volume of the culture medium to be dialysed (preferably ×50)
- Dialysis tubing with molecular weight cut-off 10 000 (for preparation see below)
- Sterile flask or bottle that may contain the above medium
- Magnetic bar and stirrer

Preparation of dialysis tubing

1. Wash tubing twice for 1 h in 1l of 50% ethanol.
2. Wash tubing twice for 1 h in 10 mM $NaHCO_3$.
3. Wash tubing twice for 1 h in 1 mM EDTA.
4. Wash tubing twice for 1 h in distilled water.
5. Store in distilled water with 10% acetic acid at 4°C.

Detoxification

1. Make two knots in one end of the dialysis tubing and boil it in distilled water for 2 h. Cool before use. All work from this stage should be sterile.
2. Remove the cell-free medium from the cell culture, leaving a minimum amount of medium for the cells to survive for approximately 2 h. If the cells are not adherent, or immobilized by other means, they may be removed by, for example, centrifugation.
3. Take the tubing in one hand with sterile gloves, pipette the medium into the tubing and immerse it in a bottle containing fresh growth medium without serum.

4. Put a magnetic bar (sterile) into the bottle and place it on a magnetic stirrer at room temperature for 2 h.
5. With a pipette, withdraw the medium from the tubing and return it to the cell culture from which it was taken.
6. If sterility is not guaranteed, the medium must be filtered through a sterile filter (0.2 µm) before returning it to the cells.

Alternative Procedure: Detoxification by Gel Filtration

Materials and equipment

- Column packed with Sephadex G-25, bed volume × 3 of the volume of the medium to be treated
- Growth medium with all supplements ***except*** serum, at twice the bed volume of the column
- Equipment for running the medium through the column and collecting the eluted fractions (pump, UV monitor, recorder and fraction collector)
- Sterile filter (0.2 µm)

Detoxification

1. Equilibrate the column with one bed volume growth medium without serum.
2. Remove the cell-free medium from the culture, leaving a minimum amount of medium for the cells to survive for approximately 2 h. If the cells are not adherent, or immobilized by other means, they must be removed by, for example, centrifugation.
3. Apply the cell-free medium to the column and start the recorder and fraction collector.
4. Elute growth medium without serum (one bed volume) and collect suitable fractions (e.g. 1/10 to 1/5 of the volume applied to the column).
5. Pool the fractions containing the proteins (peak adsorption at 280 nm) to give a total volume equal to the volume applied to the column.
6. Filter the medium through a sterile filter (0.2 µm) and return to the cell culture that it was taken from.
7. Wash the column with suitable buffer (e.g. PBS); if the column is kept, the buffer must contain a preservative (e.g. thimerosal or sodium azide).

Background information

Both the described methods are based on the observation that the toxic metabolites are usually low molecular weight compounds. The rationale is to remove these low molecular weight compounds and retain the macromolecular fraction containing growth factors and potential cell product. Several methods have been employed to remove toxic metabolites. In addition to the two methods described above, ion exchange chromatography has been used. Another approach is to detoxify the medium by adding complex binding substances. Serum appears to protect against toxic effects because low-serum cultures are often more sensitive to toxic metabolites than cultures with higher serum concentration. Albumin often functions as a complex-forming agent; however, at present no other detoxicants are known, and one has to use physicochemical methods in order to remove the toxic metabolites.

An alternative to batch treatment, such as the two methods described, is continuous removal by dialysis. Anchorage-independent cells may be grown inside dialysis tubing in roller bottles with cell-free medium outside the tubing, or in hollow-fibre cartridges. When larger volumes of medium are to be dialysed (e.g. in connection with a bioreactor), ultrafiltration membranes can be used. The medium (with or without cells) is circulated through the ultrafiltration unit and dialysed against fresh medium, usually lacking serum, which is circulated on the other side of the membrane.

Time considerations

Detoxification of the cell culture medium with one of the above-described methods takes 1-2 h. This should be taken into consideration when planning the experiments and care should be taken over storage of the cells whilst treating the medium.

OXYGENATION

The control and measurement of oxygen concentration in culture media are essential elements in *in vitro* cell culture. Oxygen is often the first component to become limiting at high cell densities because of the low solubility of oxygen in aqueous media. In addition, the supply of oxygen may also limit the productivity in cell culture. Therefore, the ability to deliver sufficient oxygen to the culture must be incorporated in a successful experimental procedure. The dissolved oxygen concentration must also be controlled, because high oxygen concentrations may result in cellular damage through the generation of oxygen free radicals. The rate of cellular oxygen consumption can be used to provide information on the status of a culture; it correlates well with the cell density in bioreactors as well as with protein production. The oxygen consumption rate can also serve as a rapid and sensitive indicator for problems developing in the culture because a rapid decrease can occur in response to changes in pH or temperature or depletion of essential medium components.

Units

The units used for oxygen concentration in media are: percentage air saturation, percentage dissolved oxygen, oxygen tension, P_{O2} or molar units. Percentage air saturation (identical to percentage dissolved oxygen, % DO) is represented by:

$$\%\text{ air sat. }(\%\text{DO})=(C_L / C_{sat})\times 100\%$$

where C_L is the actual oxygen concentration in the bulk medium and C_{sat} is the oxygen concentration in equilibrium with air (21% oxygen). Oxygen tension or P_{O2} (mmHg) is the partial pressure of oxygen in the liquid phase in equilibrium with the partial pressure of oxygen in the gas phase. At equilibrium, the partial pressure of oxygen in the liquid phase (corrected for the partial pressure of water) is equal to the partial pressure of oxygen in the gas phase. Thus, P_{O2} = 0.21(760 − 47 mmHg) = 150 mmHg at 100% air saturation, where the partial pressure of water is 47 mmHg at 37°C. The oxygen concentration in culture medium (RPMI 1.640) in equilibrium with air has been reported to be 0.2 mM and with water-saturated air to be 0.19 mM. Oxygen concentrations for other oxygen tensions are plotted. The equilibrium oxygen concentration in media can also be estimated by correcting the solubility of oxygen in water for the presence of electrolytes. Chemical methods to determine the oxygen concentration.

Effect of oxygen concentration on cells

Oxygen, while it is required for growth, can be toxic at high concentrations. Effects of varying oxygen concentrations (reported

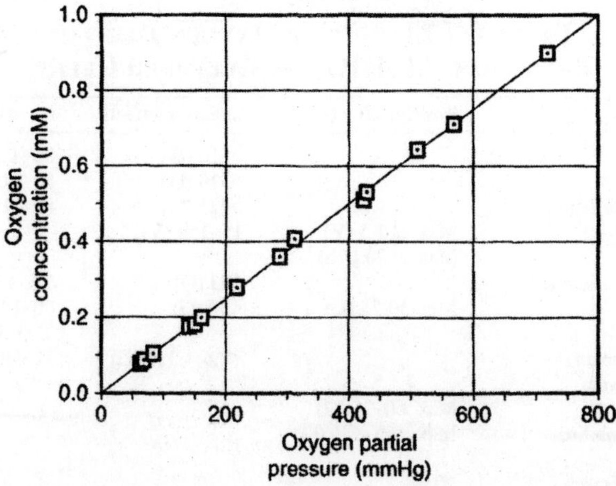

Fig. 8.15 Equilibrium concentration (mM) of dissolved oxygen in RPMI medium at different oxygen partial pressures (mmHg).

as partial pressure of oxygen, mmHg) on various properties of cultured cells are presented. Established cell lines appear to function well over a wide range of oxygen tensions (~15–90% DO). Primary cell types, however, appear to survive best at lower oxygen tensions, which better simulate the *in vivo* environment.

Typical values of the specific oxygen consumption rate

The specific oxygen consumption rate, *q,* has been reported to be affected by cell type, cell density. Proliferative state of the culture, glucose concentration and glutamine concentration. The specific oxygen consumption rate is fairly constant over a wide range of dissolved oxygen concentrations, but was found to decrease with P_{O2} below 5–10 mmHg. Oxygen consumption rates for a variety of cell types are listed in Table.

PROCEDURE; MEASUREMENT OF OXYGEN TRANSFER COEFFICIENT AND OXYGEN UPTAKE RATE

Values for $K_L a$ (in units of inverse time), which is the product of K_L, the overall mass transfer coefficient, and *a,* the gas-liquid interfacial area

Table 8.7 Effects of Oxygen Partial Pressures (MMHG) on Cultured Cells

Cell type	Positive effects	negative effects	No effect
TK6		>510 (G)	80–425 (G)
Vero		>310 (G)	80–215 (G)
SP210 hybridoma		>425 (G)	80–310 (6)
	Max. at 1.5 (V)	15–150 (V)	
	Max. at 75 (Ab)		
Murine hybridoma		143 (G)	7.5–135 (G)
	Min. 30-75 (D)	<30 (G)	30–120 (G)
			22.5–150 (Ab)
Murinelhuman hybridoma		<1.6, >320 (G)	12–160 (G)
Porcine aorta	Max. 60–75 (G)		
Human umbilical vein	Max. 105–120 (G)		
Rat hepatocytes	90 (v)		
Human haemopoietic	Gas phase 37.5 (V)	Gas phase 150 (V)	
Human macrophage progenitor	Gas phase 37.5 (V)	Gas phase 150 (V)	
Murine haemopoietic	Gas phase 3-5.25 (V)	Gas phase 7.5 (V)	
Blood rnonocytes	Gas phase 9 (V)	Gas phase 315 (V)	

Table 8.8 Reported Oxygen Consumption Rates of Cells in Culture

Cell type	Specific oxygen consumption rate ($\mu M\ h^{-1}\ 10^{-6}$ cells)
CHO	0.15
BHK-21	0.2
Human foreskin (FS-4)	0.05
Human granulocytes	0.03–0.26
Murine hybridoma	0.05–0.2
	0.03–0.37
	0.110
	0.078–0.086
	0.12–0.48
Murinelhuman hybridoma	0.267–0.452
Murine myeloma	0.057
Murine macrophages	0.06–0.07
Vero	0.24

per unit of reactor liquid volume, can be determined experimentally by the 'dynamic gassing-out method'. Equation describes the dynamics of the dissolved oxygen concentration in liquid medium without cells (no oxygen consumption):

$$\frac{dC_L}{dt} = K_L a(C^* - C_L)$$

where C_L and C^* are the actual and dissolved oxygen concentration in equilibrium with the oxygen concentration in the bulk gas phase, respectively. To measure $K_L a$, the vessel is first equilibrated with nitrogen until the oxygen concentration in the medium is negligible. The oxygen concentration is then recorded as a function of time after starting the flow of the supply gas. The value of $K_L a$ is then determined by integration of Equation (with the initial condition $C_L = 0$ at time $t = 0$), which yields:

$$-K_L at = \ln(1 - \frac{C_L}{C^*})$$

where C_L is the recorded output and C^* is the final equilibrium output. Using Equation, the slope of the line In $[1 - (C_L/C^*)]$ versus time is numerically equal to $-K_L a$. Measurements of $K_L a$ need to be performed under conditions identical to those of the culture, because $K_L a$ is a function of the method of oxygenation (i.e. surface aeration, direct sparging, etc.), liquid volume, power input, the ionic strength and viscosity of the medium, temperature as well as antifoam concentration. The equilibrium concentration of oxygen, C^*, is also a function of the ionic strength of the medium, as well as the culture temperature and partial pressure of oxygen in the supply gas. Although this method is simple and convenient, it contains some inherent sources of error due to the oxygen probe response time (time required for the output to measure a percentage of a step change). Ideally, the oxygen probe response time, τ_p, should be much smaller than the mass transfer response time, $1/K_L a$. This condition would apply for most animal cell bioreactors where a typical Clark-type (polarographic) electrode has a 95% response time of ~15 s, which is much smaller than most of the inverses of $K_L a$ values reported in Table. Other sources of error include the development of liquid films about the probe (important in viscous media or with slow agitation), and as a result of an initial variation in the concentration of oxygen in the gas phase (C^*) due to the initial concentration of nitrogen in the gas space (important in highly agitated sparged reactors where the main gas residence times are long). Where applicable, there are several papers that review corrections to the dynamic gassing-out procedure.

Once K_La is known, the cellular oxygen consumption rate, qX, can be measured under culture conditions. The mass balance for oxygen in a batch reactor with cells is given by:

$$\frac{dC_L}{dt} = K_La(C^* - C_L) - qX$$

where X is the cell concentration and q is the specific oxygen consumption rate (μmol O_2 consumed h^{-1} $cell^{-1}$). In a bioreactor with DO control ($dC_L dt = 0$), Equation can be rearranged to yield:

$$q = \frac{K_La(C^* - C_L)}{X}$$

In reactors without DO control, a value for dC_L/dt can be determined from a C_L versus t plot.

It is possible to construct a small vessel (respirometer, ~2-50 ml) for the specific purpose of measuring the cellular oxygen consumption rate. Equipment descriptions and/or procedures to measure q. The vessel is typically a glass-jacketed vessel with two ports, one for a DO probe and the other for sample injection. A magnetically coupled flea (small stirring bar) can be used to suspend the cells. After injecting the cell sample, the vessel is sealed (ground-glass joints work well) and the decrease in DO is recorded over a 5–10-min period. The slope of the DO profile is the consumption rate, because Equation reduces to:

$$\frac{dC_L}{dt} = -qX$$

for a vessel where the oxygen transfer term can be neglected. The most critical step is the removal of all air bubbles from the vessel. The value of q measured by the respirometer is independent of K_La, which was needed to determine q by Equation.

Supplementary Procedure: Oxygenation Methods

The concentration below which oxygen becomes limiting to the culture is known as the critical oxygen concentration. Below this level, the cells may be limited by energy (ATP) supply and rely more on glycolysis. Thus the method of oxygen delivery must be able to maintain oxygen levels above the critical oxygen concentration. The oxygen balance over the bioreactor implies that oxygen transfer can be increased by increasing the mass transfer coefficient K_L, increasing the area available for mass transfer, a, or by increasing the driving force for oxygen transfer, $C^* - C_L$. The methods used to oxygenate laboratory-scale reactors include surface aeration, membrane aeration and direct gas sparging. For larger and production-scale bioreactors, direct gas sparging is necessary. Overall mass transfer coefficients for these various aeration systems are listed.

Surface aeration

Surface aeration is the simplest form and is sufficient for most laboratory applications. However, this is limited to laboratory-size reactors and becomes inadequate for large systems because the surface area for oxygen transfer per unit reactor volume decreases with increasing volume. For example, $K_L a$ was found to have an approximately inverse square relationship with culture volume. Oxygen transfer by surface aeration is limited by the degree of mixing at the gas/liquid interface. Thus, increasing agitation intensity and the use of a larger impeller have both been shown to increase $K_L a$. found increases in $K_L a$ to be proportional to increases in agitation intensity. Unfortunately, there are upper limits to the level of agitation that can be used in bioreactors due to the fragility of mammalian cells. The position of the impeller in the vessel can also influence $K_L a$: $K_L a$ was essentially constant with impeller position in a 500-ml spinner flask until the impeller was

TABLE 8.9 OVERALL TRANSFER COEFFICIENTS FOR OXYGENATION USING DIFFERENT AERATION SYSTEMS

Aeration system	$K_L a (s^{-1})$
Static liquid surface	10^{-6}
Stirred liquid surface	10^{-6}–4×10^{-3}
Silastic membrane aeration	10^{-4} – 10^{-2}
Sparged with agitation	7×10^{-3}–2.5
Gas exchange impeller	4×10^{-3} – 3×10^{-2}

positioned within 1 cm of the liquid surface; at that point $K_L a$ increases dramatically (approximately four times the submerged value). Positioning the impeller near the liquid surface, however, would result in poor mixing even in most laboratory-size vessels. Thus, a second impeller could be attached to the impeller shaft and positioned at the liquid surface. This partially submerged second impeller, known as a surface aerator, was seen to increase the mass transfer coefficient, K_L, from 6.4 to 26.2 cm h^{-1} in a 1-1 bioreactor and from 3.3 to 13.5 cm h^{-1} in an 8-1 bioreactor by increasing mixing at the air/liquid interface. Oxygenation with air is difficult because the optimal DO levels (15-90% air sat.) are close to the equilibrium concentration of oxygen in the medium. Thus, the driving force for oxygen transfer ($C^* - C_L$) is small but this can be increased by supplementing with pure oxygen. In those cases where a lower oxygen concentration is desired, the driving force for oxygen transfer can be decreased by mixing the supply air with nitrogen.

Sparger aeration

Oxygen can be supplied to the culture by directly sparging the supply gas into the bioreactor. The basic concepts describing mass transfer in gas-sparged bioreactors are reviewed. Sparger aeration offers the advantages of very high oxygen transfer rates because of the large interfacial area, *a*, for bubbles. Disadvantages arise from the fact that animal cells are susceptible to damage from direct sparging and foaming of the culture medium. Nevertheless, many different cell types have been successfully grown in air-sparged bioreactors. There are two main types of sparged reactors: bubble columns and sparged reactors with agitation.

Bubble column bioreactors

The attachment of cells to bubbles and the related shear forces associated with the rise of bubbles through the culture medium are not detrimental to cells; nor is there significant cell death at the sparger site. Cell damage in sparged reactors is thought to occur primarily at the top of the reactor due to bubble break-up or shearing in thin films of collapsing foams. Thus, parameters that affect cell death are those that affect the exposure of cells to bursting bubbles and unstable foams, as well as the frequency of bubble bursts. The height of the sparged reactor affects cell viability; increasing the column height reduces cell damage due to the decrease in the surface area for bubble disengagement per unit reactor volume. The maximum viable cell density approached that achieved in surface-aerated and agitated reactors, as the liquid height/column diameter ratio exceeded 10:1, Increased column height also increases the bubble residence time in the reactor. While high gas flow rates are desirable, because $K_L a$ is proportional to the ratio of superficial gas velocity to bubble size, there is a direct relationship between increasing the specific gas flow rate (increasing bubble frequency) and increased cell death in bubble columns at a fixed column height. At present, there is no definite correlation between bubble size and cell death. Found cell damage to be independent of bubble diameter for bubbles larger than 2 mm, while studied bubble diameters less than 1.68 mm and found cells to be more sensitive to smaller bubble sizes (0.98 and 0.2 mm).

It is suggested that bubble columns are operated with large height/diameter ratios (>5) to decrease the exposure of the cells to bursting bubbles. The gas flow rate should be kept to the minimum level allowable. However, the gas flow rate needs to be sufficiently large to keep the cells in suspension and to maintain an adequate DO level. If it is not possible to keep the gas flow rates below levels at which there iş significant cell damage, the effect of smaller bubble sizes to increase oxygen transfer should be tested. The bubble diameter is affected by the pore size of the air-bubble distributor, the liquid properties of the medium and gas flow rates (only at higher gas flow rates). Perforated meta! plates can be used to achieve larger bubble sizes (4-6 mm) while sintered glass niters can be used to achieve smaller diameters. Cell viability in sparged reactors is also affected by medium components. 'Shear protectants' are commonly used in both stirred-tank reactors and in gas-sparged vessels to protect cells. For example, at high air sparger rates, the non-ionic surfactant Pluronic F-6S (BASF Corp., Parsippany, NJ) at concentrations in the range 0.1-0.2% (w/v) had a strong protective effect on hybridoma cells. To prevent foam formation in sparged bioreactors, silicon antifoams have been used at concentrations of 6-100 ppm; and in this concentration range there was little effect on cell growth.

Sparged bioreactors with agitation

Mechanical agitation is necessary for suspending the cells, circulating the medium and increasing oxygen transfer rates. Thus, lower gas velocities are more necessary in sparged bioreactors with agitation than in bubble column reactors. Reported values for gas flow rates of 1 VVh (volume gas per volume reactor per hour) for bubble columns and 0.1-1 VVh for sparged reactors with mechanical agitation. These lower gas flow rates may reduce the level of antifoams and shear protectants required.

The agitator can be located above the sparger, but it may be desirable to place it under the sparger because there is some evidence that this results in lower shear forces. In addition to perforated metal plates or sintered glass filters, gas-exchange impellers and hollow-fibre membranes have been used as air-bubble distributors. Gas exchange impellers sparge the supply gas through a steel mesh screen attached to a hollow agitation shaft. In contrast to bubble column reactors, in this system the height/diameter ratio is not relevant in minimizing injury from shear forces. Bubbles are retained longer due to agitation-induced fluid motion. However, to minimize hydro-dynamic cell damage, it is necessary to reduce or eliminate vortex formation around the impeller shaft, minimize bubble entrainment and, finally, minimize or eliminate bubble break-up at the free liquid surface. The latter can be accomplished by operating the reactor without a gas headspace and allowing bubbles to migrate through a medium-filled tube attached to a port at the reactor headplate. The outlet of the bubble disengagement tube should be connected to a sterile glass vessel to allow for the rise of a foam layer inside the disengagement tube.

Microcarrier cultures

There have been few reports of anchorage-dependent cells being cultured on microcarriers in sparged reactors. Cells grown on the surface of microcarriers are likely to be more susceptible to bubble-associated damage. In particular, foam formation is detrimental because microcarriers concentrate in the foam layer and deposit on the vessel wail due to bubble bursting. Even with antifoam (Medical Emulsion AF, Dow Corning, Michigan), the concentration of CHO cells in a sparged microcarrier reactor at low superficial gas velocities (0.01 cm s^{-1}) was ~50% lower than that in surface-aerated cultures (also using antifoam).

The culture of cells on microcarriers in sparged reactors requires a conservative approach, including low gas flow rales, small bubble diameter for high oxygen transfer, low-intensity mechanical agitation and the use of antifoams. In addition, it may prove beneficial to use 'shear protectants' (such as Pluronic F-68 or polyethylene glycols

(Papoutsakis, 1991b) to minimize microcarrier flotation (i.e. aggregation of microcarriers at the free liquid surface due to attachment to bubbles). It has been found recently that these additives reduce the attachment of freely suspended cells to bubbles, and this may also be the case for micro-carriers.

Membrane aeration

Bubble-free oxygenation by membrane aeration provides efficient oxygen transfer with minimal shear damage and minimal foaming. Design equations for membrane aeration are presented. In stirred-tank reactors, the oxygen supply gas is pumped through gas-permeable, hydrophobic, autoclavable tubing (silicon, polypropylene, polytetrafluoroethylene) wound about a fixed wire cage or a modified agitator. The main factors that affect oxygen transfer are the oxygen concentration in the supply gas, the membrane porosity and the surface area of the tubing, as well as the agitation rate because oxygen transfer occurs through convection. The reactor oxygen concentration can be controlled by adjusting the mixtures of O_2, N_2 and CO_2 (for pH control) in the supply gas. The tubing should be connected to the reactor through a sterile filter prior to the inlet and through a sterile trap at the outlet. The system should be tested for gas leaks both before and after it is autoclaved. Membrane aeration has limitations in large reactor systems, primarily due to the design complexity and the large membrane area required. He used 2.5-3.0 m of tubing per litre of medium for a 100-l reactor, thus using 250-300 m of tubing. They found that increased tubing length increases the oxygen transfer rate only up to a certain length, and above this oxygen transfer was not significantly affected. The 'critical length' of tubing needed decreases with smaller tubing diameter. It was also found that dividing the tubing into parallel segments supplied by a gas distributor improved oxygen transfer. The difficulty involved in cleaning and sterilizing membrane reactors between runs in an additional disadvantage.

Mixing

Mixing is required in animal cell bioreactors to provide a homogeneous environment throughout the bioreactor and to increase the mass transfer rates of oxygen and other nutrients to the cells. It is well known that intense agitation can be detrimental to mammalian cells in culture. The current understanding of the damage mechanisms for both attached and suspended cells in bioreactors has been reviewed recently.

Assessing Cell Damage

There are several assays available to quantitate cell death in bioreactors. The simplest assay uses Trypan blue dye exclusion from intact cells to determine live and dead cell concentrations. Cells that have died and lysed, however, cannot be accounted for using this assay.

Cell lysis, as determined by the concentration of intracellular proteins present in the culture supernatant, is a valuable indicator of cell damage due to excessive agitation. The cytoplasmic enzyme lactate dehydrogenase (LDH) is frequently used as a measure of cell damage for both suspension and attached cells in a variety of culture systems. Lactate dehydrogenase activity as a measure of cell injury should be used with care, however, because several problems may complicate interpretation of the results. The concentration of LDH in the culture supernatant can very likely be correlated with cell damage because intracellular LDH levels are constant during the exponential growth phase. However, intracellular LDH has been seen to decline towards the end of batch culture. Culture conditions may also affect the level of intracellular LDH. For example, the activity of LDH has been shown to increase in oxygen-limited cultures. In one culture system, LDH was found to be relatively stable in the culture supernatant, with the loss in activity not exceeding 5% per day. The generality of this finding has not yet been established.

Cell damage under different agitation conditions can be assessed by monitoring the increase in LDH release. To equate the LDH content in culture supernatant to the number of cells that have lysed, a standard curve must be established for each cell type and (where applicable) for different environmental conditions to which the cells were exposed. The procedures for the Trypan blue dye exclusion assay and the LDH assay are described.

Parameters used to Correlate Cell Damage Due to Agitation and/or Air Sparging

Literature reports have used the following reactor parameters to correlate the effects of 'agitation intensity' with cell injury in bioreactors: agitator rpm, impeller tip speed, integrated shear factor and Koimogorov eddy size. Additional parameters have been used for microcarrier bioreactors.

All correlations of cell injury with a bioreactor parameter should be used only qualitatively. These correlations are, at present, indicative of various trends or mechanistic hypotheses and should not be used for quantitative bioreactor scale-up. In addition, such correlations are applicable to the specific cell type, because different cell types are likely to exhibit different responses to fluid forces.

Correlations of cell damage with 'agitator rpm' are specific to a single reactor configuration because different bioreactor designs will result in different mechanical stresses being experienced by the cells. The mechanical stresses likely to be experienced by cells will depend on the impeller design and diameter, the vessel design and diameter and thc fraction of the reactor occupied by the liquid phase, in addition to the impeller rpm. Thus, results in terms of impeller rpm are relevant only for reporting trends that can be generalized to different reactor types (i.e. the protective effect of medium additives, damaging effects of certain impellers, etc.).

The 'integrated shear factor' (ISF), which is assumed to be (incorrectly, strictly speaking) a measure of the strength of the shear field between the impeller and the vessel wall, and may be somewhat more useful for scale-up purposes, is defined as:

$$\mathrm{ISF} = \frac{2\pi N D_i}{(D - D_i)}$$

where N is the impeller rpm, D_i is the impeller diameter and D is the vessel diameter. Cell growth was found not to correlate well with impeller tip speed ($2\mu N D_i$,) or the ISF, and thus the correlation of cell growth with ISF is not suited for incorporation into mechanistic models of hydrodynamic cell damage in agitated bioreactors. Information concerning damage in terms of impeller tip speed or ISF would not be transferable to different culture systems. Correlation of cell damage with 'Kolmogorov eddy size' represents attempts to correlate cell damage with more fundamental fluid dynamic parameters. The large eddies formed from the movement of the impeller through the medium eventually break down into eddies of progressively smaller sizes, forming a range of eddy sizes that decrease in size down to the smallest turbulent eddies in the dissipation range of the turbulent spectrum. The smallest eddies transfer their kinetic energy as thermal energy to the surrounding fluid through viscous dissipation. The characteristic size of the eddies in the viscous dissipation range of the spectrum of isotropic turbulence, the Komogorov eddies (η), in cm), can be estimated using Kolmogorov's isotropic turbulence theory:

$$\eta = (\nu^3/\varepsilon)^{1/4}$$

where ν (in $cm^2\ s^{-1}$) is the kinematic viscosity of the medium and ϵ (in $cm^2\ s^{-3}$) is the viscous dissipation rate per unit fluid mass, given by:

$$\varepsilon = P/\rho_f V$$

and:

$$P = N_p \rho_f n^3 D_i^5$$

where P is the mixing-power input (in $g\ cm^2\ s^{-3}$), N_p is the dimensionless power number, ρ_f is the fluid density (in $g\ cm^{-3}$), n is the agitation rate (rps), D_i is the impeller diameter (in cm) and V is the liquid volume (in cm^3) in which the energy dissipation takes place. A good estimate for the dissipation volume, V, is D_i^3, i.e. the volume of highest turbulence around the impeller. Thus, Equation becomes:

$$\varepsilon = N_p n^3 D_i^2$$

where N_p is a function of impeller geometry, impeller Reynolds number ($Re_i = nD_i^2/\nu$, a dimensionless number) and vessel characteristics, and will be either provided by the manufacturer of the bioreactor vessel or measured directly, and can be estimated from correlations. Correlations for N_p versus Re_i for various vessels and impeller designs are given or specifically for spinner flasks.

Cultures of Freely Suspended Cells

In most cases, cell damage to suspension cells in agitated bioreactors is primarily the result of air entrainment and bubble break-up. For example, cell damage to hybridoma cells in a 2-1 surface-aerated bioreactor was found to begin at agitation rates that promote air-bubble entrainment and bubble break-up near the bottom of the vortex. Cell damage became significant at 180-200 rpm in the 2-1 surface-aerated bioreactor when vortex formation and bubble entrainment were not controlled. However, in the absence of vortex formation and gas entrainment the 2-l bioreactors were agitated at rates up to 700 rpm without significant cell damage. Without vortex formation and bubble entrainment, found cell damage likely to occur at agitation rates where the predicted Kolmogorov eddy size approached the size of a single hybridoma cell (10-15 μm). More recent data suggest that the main source of cell injury in agitated and sparged bioreactors is bubble break-up at the free gas/liquid interface in the bioreactor. Other interactions (such as bubble coalescence and break-up in the bulk liquid or turbulent stresses in the bulk liquid) probably do not contribute significantly to cell injury (except for some very fragile cells), and can be ignored for all practical purposes until one reaches agitation intensities where the Kolmogorov eddy size becomes approximately equal to the cell size.

Consequently, for the case of surface-aerated bioreactors, the most important condition to avoid is the formation of a vortex that can lead to the entrainment of large, unstable bubbles that will break up at the free gas/liquid interface. When high agitation rates are required, baffles can be installed to decrease vortex formation and increase surface aeration and mixing. If feasible, gas entrainment and bubble break-up at the free gas/liquid interface can be eliminated by completely filling the reactor and operating it without a gas headspace. This will require that either direct sparging of air or oxygen or gas-permeable membranes (such as thin-walled silicon tubing be used for oxygenating the reactors. For direct sparging, gas has to be removed from the bioreactor, and this can be accomplished by an overflow vessel (above the bioreactor headplate) connected through one or multiple tubes to the headplate. The tubes and overflow vessel allow for controlled bubble break-up and foam dissipation without damage to the cells. Vessels with large height/diameter ratios (of at least 2-3) are desirable for culturing freely suspended cells if the gas headspace in the bioreactor cannot be eliminated completely. This is because these reactors reduce the ratio of the free liquid interface to the reactor volume, thus reducing cell injury due to bubble break-up at the free gas/liquid interface. Vessel geometry is important for effective mixing. The reactor vessel should have a hemispherical or a round bottom to eliminate corners where cells could settle out. Vessels can be modified with a hump located on the bottom ('profiled' bottom), directly underneath the impeller, to eliminate a dead spot beneath the impeller and improve mixing. Typically, pitched-blade, marine-type or hydrofoil impellers with two to four blades are used for agitation in mammalian cell bioreactors, in contrast to bacterial or yeast fermenters, which use Rushton impellers (vertical blades) for vigorous mixing. Rushton impellers may be perfectly suitable for agitation if completely filled bioreactors are used, as discussed above. For vessels with large height/diameter ratios, even high agitation rates may not provide sufficient mixing. In such cases, additional impellers can be attached to the agitation shaft to increase the number of mixing zones inside the vessel, resulting in increased mixing at lower agitation rates. Helical ribbon impellers have been used successfully for agitation of highly shear-sensitive cultures, such as insect cells.

Small-volume cultures (or starter cultures) are commonly established in spinner flasks. Spinner flasks (e.g. Bellco, Vineland, NJ, USA) are small glass vessels (20-5000 ml) with two ports for aeration and sampling. Agitation is provided by a magnetically driven stir bar suspended by a shaft. The stir bar must be suspended in the culture medium to avoid grinding the cells against the vessel bottom. The magnetic stir plates must be able to agitate the cultures at a constant speed (i.e. the agitation should not increase as the stir plate warms up and agitation should be free of pulsing) in the range of 30-350 rpm.

Shear protectants have been successfully used for the protection of cells at high agitation or mixing intensities in both agitated and bubble column or airlift bioreactors. For reviews and illustrations of impellers and other mixing devices.

Anchorage-Dependent Cells (Microcarrier Cultures)

Anchorage-dependent cells are grown in suspension culture by growing the cells attached to small microcarrier beads (100-300 μm) suspended in the culture medium by agitation. There are two types of microcarriers available: the non-porous microcarriers whereby cells are growing only on the outer surface of the beads, and porous microcarriers whereby cells are growing predominantly in the internal porous structure of the microcarrier beads.

For porous microcarriers, the essential mixing issue pertains to what is needed for good suspension of the beads and how much agitation/ mixing the beads will endure without being mechanically damaged. The cells in the porous beads are well protected from mechanical forces, and obviously no cells will be able to grow on the external surface of the beads if the agitation intensity is high. For porous microcarriers, it is desirable to use the highest possible agitation or mixing in order to increase the mass transfer of nutrient and metabolites in and out of the porous structure of the beads. The best strategy is to test the beads without cells at various agitation/mixing intensities before using them, and examine the beads microscopically for mechanical damage.

For non-porous microcarriers, in contrast to suspension cultures of anchorage-independent cells, damage to cells attached to microcarriers occurs at agitation intensities below which vortex formation and bubble entrainment and break-up can take place. Thus, the maximum agitation intensities should be much lower for microcarrier cultures than for suspension cultures. As a starting point, the agitation rate that will just keep the microcarriers in suspension should be chosen. When excessive shear damages cells, it will, eventually lead to detachment of cells from the microcarrier surface. This is an especially critical problem when cells on micro-carriers are used for prolonged recombinant protein expression in a serum-free medium that does not allow good cell growth. In such a case, it can be seen that the rate of cell detachment from confluent microcarriers in 100-ml spinner flasks increases as the agitation intensity increases. These agitation intensities are far below levels where damage was seen for a comparable cell line in suspension culture and below which bubble entrainment begins. Earlier data on cellular injury of cells in microcarrier cultures have been reviewed.

There is no substantial literature on direct sparging of non-porous microcarrier cultures. The difficulty is that the presence of bubbles induces bead flotation, i.e. attachment of beads to bubbles, and the formation of large bead-bubble aggregates that tend to rise and accumulate at the surface of the culture vessel, which is a highly undesirable characteristic. Nevertheless, it is possible slowly to sparge microcarrier cultures without undue cellular injury if suitable surfactants/antifoams (e.g. Pluronic F-68 or Medical Emulsion AF).

For non-sparged cultures of non-porous microcarriers, cellular injury is likely to be the result of three distinct mechanisms: interactions (collisions) between microcarrier beads, interactions of beads with the turbulent liquid (turbulent eddies) and interactions of beads with bioreactor internals, i.e. the impeller and probes. The first two are most likely to contribute to cellular injury, and the most significant source of cellular injury is likely to be the bead-bead interactions. Although models have been proposed and used to relate these mechanistic events to cellular injury, predictions remain uncertain and somewhat cumbersome for routine calculations. A rough estimate of conditions that may lead to agitation-induced cell injury is based on

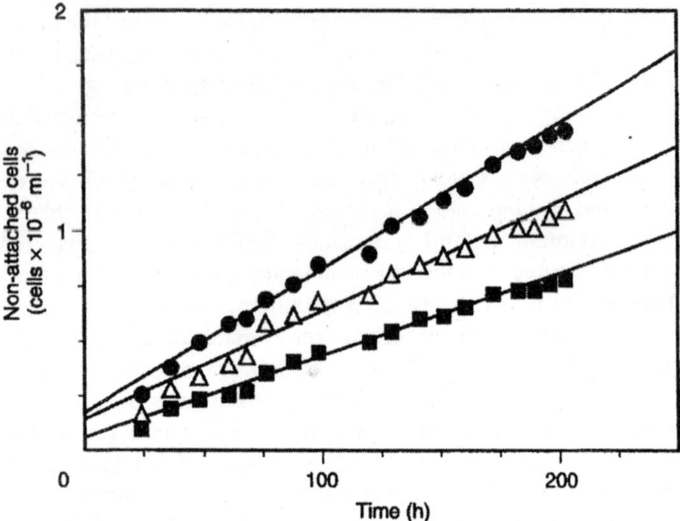

Fig. 8.16 Cumulative concentrations of non-attached CHO cells in a serum-free medium from confluent microcarriers in spinner flasks at (■) 80, (△) 115 and (●) 150 rpm. Cumulative values are the sum of suspended cells by nuclei counting in the supernatant. Microcarrier concentration was 5 g l^{-1}.

the Kolmogorov eddy length concept. Specifically, because both bead-bead collisions and bead-eddy interactions become significant when the Kolmogorov eddy length (which decreases with increasing mixing intensity) becomes similar to the bead size, the detrimental effects of agitation become quite severe as the Kolmogorov eddy lengths approach the size of the microcarriers (150-200 μm). However, it is not possible to specify precisely the smallest ratio of Kolmogorov eddy length to bead diameter (η/d) below which cellular injury becomes significant. This is because: cellular injury commences at a higher value of this ratio when higher bead concentrations are used; there is substantial error in estimating η in various reactors; and there is substantial variability in the amount of shear that various cells can tolerate. An η/d ratio of 2-2.5 (calculated according to Equations) is a good conservative estimate. Dextran is presently the only medium additive that has been shown to protect cells from mechanical damage in micro-carrier cultures by increasing the medium viscosity. When the medium viscosity is increased, the eddy length increases, and thus a culture can be agitated at higher rates without cellular injury. The beneficial effect of increased medium viscosity is more pronounced at higher agitation rates.

Beyond the aforementioned simple criterion based on the Kolmogorov eddy length, several reactor (vessel) design characteristics should be considered with respect to mixing. Selection of the proper impeller is important both for proper mixing and to minimize cell damage. The discharge and trailing vortex velocities are considerably higher from radial flow impellers (vertical blade) than from axial flow impellers (pitched blade) for identical impeller tip speeds. The higher velocities of the trailing vortexes mean higher maximum local (i.e. in the area around the impeller) shear rates, and also smaller local Kolmogorov eddy lengths. Thus, it is recommended that pitched-blade impellers be used for agitation. Spinner flasks used for microcarrier culture are typically modified with a profiled bottom (raised hump on the vessel bottom) to eliminate a dead spot where micro-carriers could accumulate. The use of a profiled bottom is more important in microcarrier culture than in suspension culture because microcarriers settle out much faster than individual cells. It is very important for the stir bar to be positioned at least 1 cm above the hump in spinner flasks to avoid grinding of the microcarriers between the stir bar and the hump. For microcarrier culture the stir drive must be able to deliver constant agitation rates, especially for the low range of agitation rates. When cells are first being attached to the microcarriers, it is critical to have slow, pulse-free agitation.

A common occurrence in microcarrier culture is the formation of large micro-carrier aggregates in which the microcarriers are joined by cellular bridges. Microcarrier aggregates made up of as many as 10 or more microcarriers are not uncommon. Microcarrier bridging occurs mainly during the growth phase of the culture, with little additional bridging occurring after cell growth has ceased. This study also showed that there is an inverse relationship between the rate of microcarrier bridging and agitation intensity. Thus, it may be of interest to operate at higher agitation intensities during the growth phase of the culture to minimize microcarrier aggregation, and to slow down the agitation as cell growth slows in order to minimize cell detachment during the later stages of the culture. In certain cases, such as to promote bead-to-bead transfer of cells to bare microcarriers, low agitation rates would be desirable during the culture growth phase.

Mechanical Protection

Mixing is required in animal cell bioreactors to provide a homogeneous environment throughout the bioreactor and to increase the mass transfer rates of oxygen and other nutrients to the cells. Intense agitation and/or aeration (air or oxygen sparging) can be detrimental to animal cells in culture. The current understanding of the damage mechanisms in bioreactors, for both freely suspended cells and cells on microcarriers, has been reviewed recently. In most studies, cell damage has been assessed on the basis of cell viability and cell death, or reduced cell growth. Although such assessments may be adequate for cell culture applications (such as in vaccine production or the production of cells for cellular immunotherapies or gene therapies) are likely to require more sophisticated assessments of the effects of fluid-mechanical forces on cultured cells. Several such effects have been investigated in the biomedical engineering literature and in the biotechnology literature. Such effects include alterations in ceil metabolism, cell cycle kinetics and DNA synthesis, alterations in protein expression and effects on the concentration of surface proteins.

Cell damage and mechanisms of protection by additives Freely suspended cells

In most practical situations, damage of freely suspended cells in agitated and/or aerated bioreactors is due to the interactions of cells with bubbles and rearranging gas/liquid interfaces. Specifically, cellular injury results from the forces released on the cells that have been collected near the bubble surface during the break-up of bubbles at the free liquid surface, or alternatively from forces released on cells entrapped in thin liquid films generated during the collapse of foams at the liquid surface. Several additives have been used since the early 1950s to protect cells from fluid-mechanical damage. Serum and the pluronic family of non-ionic surfactants are the best documented and most widely studied. Other additives include several derivatized celluloses and starches, cell-derived fractions and proteins. Additives that protect freely suspended cells from fluid-mechanical injury either decrease the fragility of the cells by nutritional or other biological mechanisms, or alter the magnitude and/or frequency of forces released on cells due to interactions with gas/liquid interfaces by physicochemical mechanisms. They have proposed the following terminology regarding the nature of the protection effect or mechanism of an additive:

- A 'biological' protection mechanism or effect implies that the additive changes the cell itself to make it more shear resistant
- A 'physical' protection, mechanism means that the cell resistance to shear remains unchanged, but that the factors that affect the level or

frequency of transmitted shear forces to the cell in a given culturing system have changed so that less cell damage is observed

- The 'biological' effect may be the result of alterations that take place either very fast and without requiring metabolic events, or requiring changes that demand metabolic events brought about by the additive. For the latter case, a longer exposure (at least a substantial fraction of the cell cycle time) to the additive would be required. The former will be referred to as a 'fast-acting biological' mechanism, and the latter as a 'metabolic biological' mechanism

Microcarrier cultures

In microcarrier cultures, cell injury is due to the interactions (collisions) between microcarrier beads on which cells are attached, interactions between microcarriers and small turbulent eddies and interactions between microcarriers and bioreactor internals, such as the impeller and various probes, To protect cells against injury in microcarrier bioreactors, an increased medium viscosity has been the only documented and studied medium alteration. Dextran (Sigma) has been used to increase the medium viscosity as discussed below. This is a purely physical mechanism of protection.

Preliminary Procedures Additive Preparation

Prepare a solution of each additive as specified by the manufacturer and/or as discussed below, filter-sterilize the solution and add to the sterile culture medium.

Procedure: Testing Before using an Additive

Before any additive is to be used with a given cell type, it should first be tested to ensure that it has no detrimental effects on cell growth, metabolism, differentiation, protein expression, etc., and that it does indeed offer mechanical protection in agitated and/or aerated systems (bioreactors).

Parallel static cultures

1. Run cultures in well plates or T-flasks (in duplicate or triplicate) with and without the additive.

2. Use an average, and the highest reported concentration, of the additive for the initial screening. If a new additive is being tested, use a range of concentrations.
3. If a small inhibitory effect is found (which is quite usual), proceed to assess whether or not the additive offers the desirable mechanical protection.
4. If the additive offers mechanical protection, check if the protection effect is present at lower non-inhibitory concentrations. Most cells can overcome a small inhibitory effect by being acclimatized to the additive.
5. To this effect, culture the cells over several (e.g. 10-20) generations in the pres ence of the additive in static cultures, and check if they can overcome the inhibitory effect. If they do, then propagate the cells (when preparing the culture inoculum) in the presence of the additive.

Check for mechanical protection effect

1. Decide what aeration and/or agitation conditions are desirable for the appli cation, and run experiments to compare cultures with and without the additive under such conditions.
2. Run the experiments in duplicate or triplicate, and repeat the experiments two or three times to ascertain the protective effect.
3. If no protective effect is found, check if protection is provided under less severe agitation and/or aeration conditions.
4. If protection is found, check how good the protective effect is by comparing with parallel static or low-agitation cultures.
5. If time and resources permit, determine the highest agitation and/or aeration conditions under which the additive will provide protection.

ADDITIVES FOR FREELY-SUSPENDED CELLS

Pluronk F-68, other pluronic polyols, polyethylene glycol (PEG), polyvlnyl alcohol (FVA) and polyvinylpyrrolidone (PVP)

Pluronic polyols (such as F-68, BASF, Parsippany, NJ) and PEG (Sigma) are the best and most widely studied and used additives, and should be the additives of choice for most, if not all, applications, Polyvinyl alcohol (Sigma) should be included in this list, although it has not been widely studied. The non-ionic surfactants Pluronic F-68 and F-88, which are block copolymer glycols of poly(oxyethylene)-poly(oxypropylene)-poly(oxyethylene), have been known to protect cells against fluid-mechanical damage in agitated and aerated bioreactors and have been used for over 30 years. These, and related, surface-active agents are used as medium additives in static, agitated and/or aerated cell cultures. More detailed studies on the effects of various other pluronics and reverse pluronics (the order of the copolymers is reversed with poly(oxyethylene) in the middle) on cell growth and/or shear protection have been published. Studies have also shown that pluronics have a concentration-dependent positive or negative effect on cell growth in 'static cultures'. Although it is possible that the effects observed could be due to small unknown impurities, these results suggest that these polyols may affect the growth of some cells independently of agitation or aeration.

Two types of polyvinylpyrrolidones (PVP-Bayer, Leverkusen, Germany) have been found beneficial to cell growth in spinner cultures, and a mixed molecular weight PVP has been found to protect hybridoma cells against shear injury in a bubble column reactor.

Polyethylene glycol (which is related in chemical structure to pluronic polyols) and PVA have been tested more recently as protectants against aeration and agitation cell damage. Both appear to offer mechanical protection as good as F-68, without any detrimental effects on the cells they have been tested on (Chinese hamster ovary (CHO) and hybridoma cells CRL 8018).

The mechanisms through which pluronic polyols, PEG and PVA protect cells from agitation and/or aeration damage have not been firmly established and have been the subject of considerable debate and speculation. Because F-68 and PEG (and also PVA) can protect cells from fluid-mechanical damage in bioreactors even after a short exposure, it is likely that their effect is either physical or fast-acting biological in nature. Some viscometric studies have shown that, unlike serum, PEG and F-68 do not protect cells against shear damage in viscometric flows (well-defined, laminar Couette flows) after prolonged (cells grown in the presence of the additive), short (30 min) or intermediate (1-4 h) exposure.

Viscometric studies are used to assess the shear robustness or fragility of cells by exposing them (usually for a short, 10-20 min, time period) to the reproducible and well-defined laminar flow of a viscometric device. In contrast, the (usually turbulent) flow in a bioreactor is not well defined and not easily reproducible, unless the same reactor is used.

These findings suggest that the shear-protective effects of these additives in the bioreactor are physical in nature, and specifically purely fluid-mechanical, i.e. due to changes in the interactions between bubbles, draining films and the cells. If their effect was biological, cells would have been protected in both shear environments (viscometer and bioreactor). Other experiments suggest that the protection mechanism may vary for different cell types.

For most applications, 0.1% of F-68, F-88 and various PEGs and PVAs should be sufficient for mechanical protection. Higher concentrations of pluronics (up to 0.2%) may be necessary in some cases, but this has not been well documented. Among the large variety of PEGs available, those of molecular weights between 1400 and 15 000 have been found successful for mechanical protection. Polyvinyl alcohol of only one molecular weight (10 000) has been examined, and has been found to have excellent protective properties.

Serum

Serum is a good but undefined shear protectant. It promotes better cell growth in agitated and/or aerated cultures in a dose-dependent fashion up to 10% (v/v). Several studies employing bioreactors and well-defined Couette flows in a viscometer have shown that the protective effect of serum is due to a physical and/or a biological mechanism. It seems that the most significant contribution comes from the physical mechanism, so any type of serum (i.e. independently of its nutritional and hormonal value) that does not inhibit cell growth can be used. For most applications, 2-5% serum will be sufficient to provide protection. Because serum complicates protein purification, may elicit several biological responses or may become a source of undesirable variability or viral contamination, its use as a protective agent against mechanical injury should be avoided. Experimental evidence suggests that the synthetic, well-defined additives (e.g. pluronics, PEG, PVA) provide as good (or almost as good) mechanical protection as serum does.

Derivatized celluloses

Derivatized celluloses are promising but not yet reliable shear protectants. They have been used since the very early days of the suspension animal cell culture technology in the 1950s. He has reviewed the early uses of methylcelluloses (MCs), presumably as protectants against shear damage for the cultivation of a large variety of suspension cells, including human, mouse and monkey cells. Also, the most detailed study on the effects of various MC grades on various cells of variable shear fragility in both shake-flask and static cultures has been carried out. Despite this work and the later work of several other investigators, MCs and other derivatized celluloses (such as carboxymethyl celluloses, CMCs) have not been established unequivocally as reliable shear protectants.

The effect of MCs and CMCs as additives to protect animal cells from shear damage in agitated, bubble column or otherwise mixed cultures may depend on cell type, MC grade and MC make. In addition, MCs and CMCs may elicit biological responses from cells, either positive or negative; these responses appear also to be dependent on cell type and additive grade and make. It is not known whether the stimulatory or inhibitory effects are due to MCs and CMCs themselves or to small impurities. The effect of some MCs in causing cell aggregation for some cells has also been reported. There is little question, however, that even after almost 40 years of research very little is understood about the mechanism and generality of the protective effect of MCs.

He found that low concentrations of low-viscosity MC were as effective as a protectant as the higher viscosity grades, thus concluding that the MC on insect cells derives from the higher medium viscosity. Higher concentrations or grades of MCs produce very viscous solutions, and some of the early work used MCs that produced media of very high viscosities.

They have reported the effect of various MCs and (hydroxy-propyl)methylcelluloses (methocels, Dow Chemical Company) and dextran in protecting insect cells (Sf9 and TN-368) in viscometric flows. All media contained 10% foetal bovine serum (FBS) in addition to the additive. They found that the best protection (58-76-fold less cell lysis compared with the unsupplemented medium) was offered by higher concentrations and higher molecular weight methocels. The best protection was offered by the additives resulting in the highest medium viscosity (4-25 times the unsupplemented medium viscosity). Similarly, dextran (MW = 476 000) offered substantial shear protection only at high concentrations (4.5%), where the medium viscosity is also high (6.6-fold higher than the unsupplemented medium). The significance of these findings in terms of protection in bioreactors is not known.

In the last 25 years, the use of MCs (in combination with other shear-protecting additives, such as serum, yeastolate or protein hydrolysates) appears to have been restricted to the cultivation of insect cells, probably because of the more widespread use of Pluronic F-68, which does not cause any of the cell-aggregation problems often associated with the use of MCs and CMCs.

They have recently reported the beneficial effect of methocel (MW 15 000, 50 000 and 100 000) on CHO cells in agitated and aerated bioreactors. This is the only recent and complete (with control experiments) reported study of derivatized celluloses as shear protectants in bioreactors. These findings appear very promising, but additional work will be needed before the widespread use of MCs as shear protectants.

Dextrans and modified starches

Dextrans and modified starches are not suitable for shear protection. He has tested a large number of synthetic polymers - including hydroxyethyl starch (HES), CMC (Edifas B50), modified gelatin (Haemaccel) and several dextrans - for improved growth of human lymphocyte and lymphoblastoid cell lines in agitated cultures with low serum concentrations. Experiments were carried out in 100- or 500-ml spinner flasks at 100 rpm. It was found that the dextrans had a positive effect on the growth of two cell lines but no effect on a third line. Both HES and CMC had a positive effect on the growth of all cells tested, but the modified gelatin had a negative effect on growth. There were no static control cultures presented to test that their agitation intensity is indeed damaging to cells.

Radiolabelled polymers were used to show that CMC and HES are not metabolized by the cells. He found that dextran of MW 488 000 offered no protection against sparging damage in a bubble column reactor. Extensive laboratory studies have established that 1-3% (w/v) dextrans (MW = 229 000) do not protect hybridoma cells from damage due to bubble entrainment and break-up in an agitated bioreactor, but instead increase cell death under intense agitation. Control static or low-agitation cultures showed that dextran does not affect the cells, at least in terms of growth rates, so its detrimental effect is due to the viscosity increase. These findings are in contrast to the viscometric studies.

In summary, the effect of dextrans can be positive or negative depending on the cell type, bioreactor or even dextran grade and make used, but all experimental evidence so far suggests that dextran offers no mechanical protection. Dextrans and modified starches *should not* be used as protectants against shear damage.

Proteins and cell extracts

For proteins and cell extracts there have been mixed results and insufficient testing. Several protein mixtures (in addition to serum) and a protein have been used as shear protectants for the cultivation of various cells. Again, their protective effect is not established unambiguously due to lack of proper control experiments. They have been reviewed in detail and include: peptone, lact-albumin hydrolysate, tryptose phosphate, yeast extract, yeastolate and Primatone RL (a peptic digest of animal tissue). Bovine serum albumin (BSA), a major component of bovine serum, is a widely used additive in serum-free media and has been used frequently as a shear protectant by several investigators; it was recently shown to be an effective additive against fluid-mechanical damage of hybridoma cells in an airlift bioreactor, although it has no effect on the cells in static or spinner cultured. Among all these additives, BSA is the only one recommended for use as a shear protectant (if it is also included as a medium additive for other reasons). Otherwise, even BSA should be avoided in favour of the well-defined synthetic additives, such as F-68, PEG, PVA and PVP.

Additives for Microcarrier Cultures

As already discussed, theory and experimental evidence suggest that an increased medium viscosity will reduce damage of cells on microcarriers because of the reduced intensity and frequency of interactions that cause cell damage in micro-carrier cultures.

Among the possible additives that can increase medium viscosity with no apparent detrimental effects on the cells, dextran (up to 0.3%, w/v) has been found beneficial to these cultures. A limited number of experiments have shown that MCs and CMCs are not suitable protective agents because they cause micro-carrier aggregation and flotation (i.e. collection on the liquid surface in the form of aggregates).

Use dextran of high molecular weight (between 200 000 and 300 000) at 0.15% or 0.2% (w/v) or, if necessary, lower molecular weight dextrans at 0.3% (w/v).